Ola Larses

Architecting and Modeling Automotive Embedded Systems

Ola Larses

Architecting and Modeling Automotive Embedded Systems

Technology, Methods and Management

VDM Verlag Dr. Müller

Impressum/Imprint (nur für Deutschland/ only for Germany)
Bibliografische Information der Deutschen Nationalbibliothek: Die Deutsche Nationalbibliothek
verzeichnet diese Publikation in der Deutschen Nationalbibliografie; detaillierte bibliografische
Daten sind im Internet über http://dnb.d-nb.de abrufbar.
Alle in diesem Buch genannten Marken und Produktnamen unterliegen warenzeichen-, marken-
oder patentrechtlichem Schutz bzw. sind Warenzeichen oder eingetragene Warenzeichen der
jeweiligen Inhaber. Die Wiedergabe von Marken, Produktnamen, Gebrauchsnamen,
Handelsnamen, Warenbezeichnungen u.s.w. in diesem Werk berechtigt auch ohne besondere
Kennzeichnung nicht zu der Annahme, dass solche Namen im Sinne der Warenzeichen- und
Markenschutzgesetzgebung als frei zu betrachten wären und daher von jedermann benutzt
werden dürften.

Coverbild: www.purestockx.com

Verlag: VDM Verlag Dr. Müller Aktiengesellschaft & Co. KG
Dudweiler Landstr. 125 a, 66123 Saarbrücken, Deutschland
Telefon +49 681 9100-698, Telefax +49 681 9100-988, Email: info@vdm-verlag.de
Zugl.: Stockholm, Royal Institute of Technology (KTH), 2005

Herstellung in Deutschland:
Schaltungsdienst Lange o.H.G., Zehrensdorfer Str. 11, D-12277 Berlin
Books on Demand GmbH, Gutenbergring 53, D-22848 Norderstedt
Reha GmbH, Dudweiler Landstr. 99, D- 66123 Saarbrücken
ISBN: 978-3-639-09870-9

Imprint (only for USA, GB)
Bibliographic information published by the Deutsche Nationalbibliothek: The Deutsche
Nationalbibliothek lists this publication in the Deutsche Nationalbibliografie; detailed
bibliographic data are available in the Internet at http://dnb.d-nb.de.
Any brand names and product names mentioned in this book are subject to trademark, brand or
patent protection and are trademarks or registered trademarks of their respective holders. The use
of brand names, product names, common names, trade names, product descriptions etc. even
without
a particular marking in this works is in no way to be construed to mean that such names may be
regarded as unrestricted in respect of trademark and brand protection legislation and could thus
be used by anyone.

Cover image: www.purestockx.com

Publisher:
VDM Verlag Dr. Müller Aktiengesellschaft & Co. KG
Dudweiler Landstr. 125 a, 66123 Saarbrücken, Germany
Phone +49 681 9100-698, Fax +49 681 9100-988, Email: info@vdm-verlag.de

Copyright © 2008 VDM Verlag Dr. Müller Aktiengesellschaft & Co. KG and licensors
All rights reserved. Saarbrücken 2008

Produced in USA and UK by:
Lightning Source Inc., 1246 Heil Quaker Blvd., La Vergne, TN 37086, USA
Lightning Source UK Ltd., Chapter House, Pitfield, Kiln Farm, Milton Keynes, MK11 3LW, GB
BookSurge, 7290 B. Investment Drive, North Charleston, SC 29418, USA
ISBN: 978-3-639-09870-9

Notes for the reader

This book provides thorough coverage of the topic of Architecting and Modeling Automotive Embedded Systems. The different chapters are to some extent independent and can be read separately by the initiated reader.

For the academic reader, chapter 1 introduces the research project and provides a motivation, a description of research questions and the applied research approach.

For the automotive engineer, and the interested layman, chapter 2 provides a broad overview of the state of the art for automotive electronics, in terms of applications, technologies and architectures. Much of this material may be familiar for a seasoned engineer but may serve as a source of reference.

For the safety analyst and quality engineer, chapter 3 provides a deeper discussion on the cost-efficiency and dependability requirements posed on automotive systems. The dependability concept and methods to achieve dependability, specifically targeting safety, reliability and maintainability, are described in detail.

For the reader interested in managing model based design section 4 provides an overview on such aspects. Chapter 6 provides more details on the modeling techniques and the models themselves.

A system architect may be interested in chapter 5 that provides a methodology for architecture design based on quantitative methods.

And for the restless manager, chapter 7 summarizing the conclusions may be enough…

…good reading,

Ola Larses

Contents

1   Introduction - The Automotive Challenge Revisited

This section introduces the topics of this thesis and provides a brief background that places the other chapters in a context. An overview of the results of the research is also included in this introduction. This thesis is the result of the CODEX research project, a background on this project is also provided together with some discussions on the adopted research methodology.

1.1 The transition to a new technology

The extensive use of electronics in modern vehicles is well known. Electronics and software provide possibilities for substantial improvements in functional content, performance and other product properties. At the same time this field of engineering has been uncharted territory in the automotive industry, the maturity in the R&D organizations to manage embedded computer systems is generally low. The engineering departments are barely coping with the increased system complexity, and maintaining quality is costly and reactive rather than proactive. For example, in 2003 it was reported that 49.2% of car breakdowns in Germany were due to Electrical/Electronics (EE) failures [Knippel & Schulz 2004]. In response, car makers have reduced the number of software based functions in the car [Auer 2004] and also made major recalls based on electrical problems [Hutton 2005].

One important challenge for the automotive industry lies in engineering electronics to achieve a competitive advantage through cost-efficient and dependable systems. The degrees of freedom created by powerful embedded control systems need to be managed and exploited.

With changing technology the content and complexity of system architectures of automotive electronics have evolved. The first systems were simple enough to be designed and maintained with a minimum of engineering, which was still the case in the 1950s. In the 1990s the complexity had grown substantially and software based electronic control units (ECU) had made an entry. The Mercedes S-class from 1991 included more than 50 ECUs and more than 3 km of wiring [Hofmann & Thurner 2001], the current BMW 5 and 7 series have 70 networked ECUs [Reichart & Haneberg 2004]. The amount of embedded code is also rapidly increasing, the Mercedes S-class carried 1MB of code in 1990, 100MB in 1998 and 500MB in 2005. Simultaneously the automotive manufacturers must meet a customer demand for more individualized vehicles introducing the need for variability. As the EE systems are continuously improved and altered both versions over time and variants selected by customers must be managed. The basic product properties of automotive EE systems are illustrated in Figure 1-1, and the special attention given to the system architecture in this thesis is indicated by the dashed box.

Figure 1-1 The product properties and requirements of automotive EE systems

1.1.1 Adopting a lifecycle perspective

If you are asked to develop a cost-efficient and dependable control system architecture for automotive applications you will inevitably begin to examine the current systems, the technologies at hand and the individual solutions. Then you will start looking for improvements and small changes, or maybe even something bold like redrawing the entire solution! Probably the results are at least good, if not impressive, and it appears as if you have fulfilled your task. However, one latent problem that may remain when you are finished is the need to continuously develop, improve and change your solution, and the question is: Will the new solution easily adapt to the requirements introduced for the next generation of the product? If it does not, then you have actually failed to be cost-efficient. To really succeed the solution must also be sustainable.

But how can you ensure a sustainable solution, with long term cost-efficiency and dependability? First of all, it must be recognized that a product has a life after the drawing board. Every effort in the development stage of a product should aim at the entire product lifecycle. Even from the earliest conceptual design decisions concerning the system architecture it must be recognized that the product will pass through a development stage, a sales-to-delivery stage and a service stage, the concerns may even be extended to the phase of disposal of the system.

Figure 1-2 A lifecycle perspective on the process

The development process entails all activities from new ideas to the point where changes are introduced in the production of the product.

The sales-to-delivery process is initiated in the contact with a customer choosing a specific configuration, linked by production and ended when both product and sufficient knowledge about the product have been transferred to the customer.

The service process entails all interaction with owners and users of the vehicle after the delivery.

The development process must be clearly separated, but also properly linked to the sales-to-delivery process and the service process. The requirements of dependability and cost-efficiency on each of the processes must be well understood. The resulting overall cost-efficiency and dependability are provided by the performance in each of these lifecycle stages, but in this thesis a specific focus is placed on the development process as indicated in Figure 1-2.

1.1.2 A systems perspective on embedded system design

The design process depends on the properties of the product. In order to understand this relation thoroughly it is useful to review some basics of systems theory. According to general systems theorists it is possible to engineer systems that consist of a large number and a small number of components [Weinberg 2001].

The law of large numbers states that: "the larger the population, the more likely we are to observe values that are close to the predicted average values". For example, consider the properties of a gas in a bottle. We do not need to look at the specific molecules, but can study average properties such as volume, pressure and temperature. Statistical mechanics deals with such systems. The theory applies to complex systems that are sufficiently random in their behavior. As a first approximation, the number of objects is a measure of complexity and randomness is the property that makes statistical calculations valid.

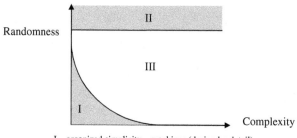

I – organized simplicity – machines (design by detail)
II – unorganized complexity – populations (design by statistics)
III – organized complexity – systems (architect)

Figure 1-3 Engineering and Architecting (Based on Weinberg [2001])

The model is illustrated in Figure 1-3. Systems with many components can be treated as populations and designed by statistics, represented as area II in Figure 1-3, while small systems can be designed by detail (area I). The problematic area is in between, represented in Figure 1-3 by area III, concerning systems too complex for analysis while being too organized for statistics. This is the field in which systems design and architecting becomes an art.

There is a lack of means to deal with systems of type III – systems of medium numbers. The Law of Medium Numbers states that *"for medium-number systems, we can expect that large fluctuations, irregularities and discrepancies with any theory will occur more or less regularly."* [Weinberg 2001]. (Anything that can happen, will happen.) System theory came about as knowledge moved from the mechanical view (I) to the organized complexity world (III). If a machine becomes complex enough it must be designed utilizing system principles. Software intensive systems are often of such nature and the increased reliance on software in automotive systems is pushing them into the same field.

One kind of system principles is to use heuristics and hope for the best. However this is not good enough for automotive applications, where large fluctuations and irregularities can not be tolerated. Automotive systems must at some point in the development process be pushed to the machine region (I) so they become analyzable. This can be achieved by utilizing the system principle of architecture design and decomposition, another option is to make extensive use of computer support.

1.1.3 Architecting and systems engineering

One way to deal with complexity is decomposition of a system into simpler parts that are analyzable (each part belongs to region I in Figure 1-3). Designing an architecture is to define the structure of major building blocks in a system, applying abstraction and decomposition and thereby defining parts and connections. The system can be analyzed as a collection of parts together with the relationships between them. However, in many cases the problem is pushed from the parts to the connections between them that still are complex. The connections can be synthesized into a new field of knowledge such as electromagnetism, physical chemistry, etc. Providing a system architecture for automotive applications is in line with this reasoning where the protocols on the serial buses (e.g. the CAN communication) have turned into a "component".

Looking at the architecture of an evolving system with a lifecycle perspective it is not possible to predict and propose a technical solution that is sustainable over time. An architecture is necessary but not sufficient. Therefore, an architecture design process is needed, as well as an architecture applicable for the current requirements.

With growing EE (electrical/electronic) systems, choosing and maintaining an architecture is an important issue for the future. Good design will improve the capability to build advanced functionality, and the architecture of the on-board network and choice of related components also has an increasing leverage on costs. The system architecture is an important part of the complex and configurable product.

1.1.4 Computer Aided Engineering

Another tool when dealing with complexity is computer support. With modern computers that can handle and link large amounts of information the field of organized simplicity (I) in Figure 1-3 is growing. Today it is possible to systematically design more and more complex systems by utilizing the computing power of modern computers.

With codification of product data good support can also be provided for organic system growth with less emphasis on efforts to provide an optimal decomposition in the architecture of the implementation. Such support is scarce today but is increasingly used. This may be necessary considering the extensive concurrency in the development of more advanced automotive EE systems, reducing the possibility to plan your architecture. Tool support for organic growth places high requirements on the tool architecture dealing with the product data and also on the tools themselves to interact with the data infrastructure. It is very important that such systems provide support for the entire product lifecycle in order to ensure that systems can be maintained and repaired.

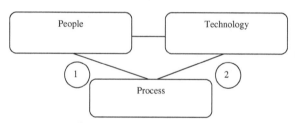

Figure 1-4 People, process, technology framework [Nambisan & Wilemon 2000]

Nambisan and Wilemon [2000] propose a framework where measures to improve product development are viewed in three dimensions: people, process and technology, see Figure 1-4. They discuss how the field of research that they refer to as New Product Development research has focused on issues related to People and Process (social interaction) indicated by ① in the figure, and research in Software Development has focused on Technology and Process (formalization of processes) indicated by ② in the figure.

It is important to recognize that both people and technology must be organized to support a process, as shown in Figure 1-5. A development approach supported by computer based tools must also be accompanied with people that are fit to work in such an environment. Similarily tools must be applied and adapted to support the needs of the people. The people aspect includes issues like team building and corporate culture, aspects often overseen in software related research [Nambisan & Wilemon 2000].

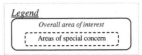

Figure 1-5 Organizational process support

With this background, tools and technology as well as human resources are illustrated as a subset of the organization in Figure 1-5. The special concern for model based product data management is indicated by the dashed box in the figure. Model based product data management is a special application of model based development.

A definition of Model based development

Models have always been used in engineering. There are several definitions of what a model is. The concept of model can be defined as:

A model is a simplified representation of a real or imagined system that brings out the essential nature of this system with respect to one or more explicit purposes.

Models are used implicitly in the mindset of the engineer, in terms of construction of physical models/prototypes, in terms of symbolic models such as written documents, and since the advent of computers by the use of CAE tools. Symbolic models are the main focus here. Symbolic models can be further classified regarding their degree of formality, that is, the degree to which they can be unambiguously defined by mathematics. Block diagrams without well defined semantics are an example of *non-formal* symbolic models. *Formal symbolic models*, which might also be called mathematical or *analytical models*, have shown to be very important tools for clarifying and solving engineering problems. Examples of models rigorously defined by mathematics include geometrical descriptions in CAD and transfer functions used to describe input-output behavior of dynamic systems.

There are many definitions of model based development[1] (MBD), here it refers to:

development based on abstract representations with predefined and documented syntax and semantics, supported by tools.

Model based approaches are further developed in chapter 4 and 6.

---

[1] Other terminology also flourish, examples include Model Driven Engineering (MDE) and Model Driven Development (MDD).

Model based architecture engineering

The architecting process is based on trading system properties against each other. A system usually has a range of contradicting requirements that must be balanced in the design. Architecture design is often referred to as an art, performed in the conceptual stages of a design process [Rechtin & Maier 1997]. Applying computer support, the architecting becomes less of an art and more of an engineering discipline. Further, considering the number and change rate of systems, performing architecture design based on codified data would be beneficial as work may be automated or reused, supported by computers. The computer support is based on product modeling and facilitates support for configuration management and product data management, applying a lifecycle perspective.

1.1.5 Aligning Product, Process and Organization

The changing technologies and product architectures require new processes and new organizations. According to Eppinger and Salminen [2001] the three domains product, process, and organization must be considered when managing complex product development. It is expected that industrial firms in which the interaction patterns across the three domains are well aligned will outperform firms for which the patterns are not aligned. The model is illustrated in Figure 1-6.

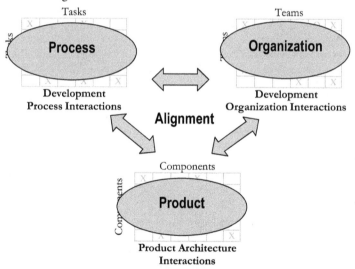

Figure 1-6 The Eppinger-Salminen framework

The interaction patterns are found by decomposing each of the three domains into interacting entities. The product is decomposed into subsystems and components (similar to the product architecture). The process is decomposed into sub-processes, tasks and activities. The organization is decomposed into teams and possibly working groups and individuals. The interaction patterns in the different dimensions should be strongly related as the development organization is executing the development process, which is implementing the product architecture [Eppinger & Salminen 2001].

In a purely mechanical product where the interactions between components are defined by how they are physically interconnected it is possible to look at the product and "see" the organization and required processes. This is also the traditional way of organizing automotive companies. It is well known and often mentioned that the organization in automotive companies is a replication of the product architecture. Recently the organizations have been changing in response to the increased content of electronics and software. Managers of EE systems have gained rank and the people working with software and electronics have been broken out of the traditional organization structure.

Adopting the trinity of Eppinger and Salminen it is assumed that if product complexity (compare product architecture interactions) is increasing, work must be done in both the process domain and in the organization domain to maintain alignment and thereby outperform the competition. The product properties illustrated in Figure 1-1 map to the product dimension in the model, and meeting the requirements in the design is a problem for the development process and organization.

As all three perspectives are to be aligned it is possible to choose where to perform the most changes and what perspective that should be allowed to develop organically with less management control. Trying to maintain parts of the organization and development process it is possible to change the product architecture to achieve alignment. This entails that the meaning of components is redefined and that the components are arranged in a pattern that suits the organization and process.

It must also be considered if organizational changes take precedence over process changes or the other way around. Organizational efforts such as matrix and project organization structures can be used to align to complex product interaction patterns while processes are adapted to the organization. Alternatively can process efforts, such as the introduction of development tools and formalized development models, manage interactions and the organization is adapted to the process.

It is expected that combinations of changes are needed, adapting the architecture in line with the existing organization and process, and simultaneously changing the organization and process to fit the new technology. Balancing these changes is a delicate management issue.

Given that there are optional approaches to develop and integrate complex products, as indicated by the Eppinger-Salminen model, it is possible to identify at least two options. One approach is based on architecting and organizational changes with less focus on the process; this approach requires extensive social interaction in the development and formalized ways of decomposing work tasks in the organization in line with the product architecture. Another approach enhances the development process through formalization of product data management and provides computer based support for this process, allowing the product to grow more organically with less concern for the mapping of the product architecture to the organization. Management of product complexity can thus be achieved either in the product by architecture design or by the formalization of the product data management in the design process. This topic is further discussed in chapter 4.

Combining these two options would provide an approach that focuses on the architecture and also provides a formalized process for this work, moving from the art of architecture design to architecture engineering. The two approaches are the basis for the structure of this thesis and both aspects are further developed.

Further, the organization and process as well as the product architecture must cope with continuous and sustainable development. Over a lifecycle a product is not only developed and improved, but also produced and maintained in a process performed by an organization. Through further reasoning in line with the previous ideas, one approach to deal with the lifecycle issues is to provide a system architecture that enables continuous development through modularity and scalability. Another approach is to formalize the development process and codify product data in a way that makes the properties of the system transparent and improves the possibility to perform efficient system analysis and synthesis, also supporting the sales to delivery and service phases. Analogue to the previous discussion, combining the two efforts should be the most effective solution, however not necessarily the most cost-efficient unless the efforts can be balanced properly.

1.2 Contributions and Scope of the Thesis

This section relates the thesis to other sources and domains and establishes the scope and relevance of the work.

1.2.1 Contribution

This thesis provides a broad overview of the current situation in the field of automotive embedded systems, covering challenges and solutions from both an academic and industrial perspective. The industry requirements on cost-efficiency and dependability are defined, decomposed and interpreted. The requirements are related both to architectural alternatives and design methods. The thesis provides a thorough review of cost and dependability related topics and proposed state-of-the-art solutions. A thorough review of the current situation in the automotive industry in terms of technologies and industrial initiatives, the state-of-practice, is also provided.

Another contribution of the work is the architecture design method for modular embedded automotive systems. The method has been developed based on known theories and applied in automotive case studies. In the process of implementing the architecture design method product data management related problems were discovered and a simple information model was developed to understand these problems. Extensions of the information model utilized for architecture design to serve other purposes are discussed and proposed.

The proposed information model provides a basis for the analyzable documentation needed to ensure the dependability of the system, specifically in terms of the need for reliability, maintainability and safety. Utilizing an information model provides traceability within the product across components, and also between different organizational units using different views of the product throughout the lifecycle process. A more formalized documentation process, possibly relying on databases, is seen as a manifestation of a model based development process. Management guidelines for the degree of formalization and application of model based principles are also proposed. Further, the context and some general properties of models and model based development are discussed for a broader understanding of the topic.

The collective work applies the Eppinger-Salminen framework in the automotive industry with a special focus on embedded control systems, and provides the framework with domain specific content.

1.2.2 Outline of the thesis

Placing the changing automotive electronics, the lifecycle perspective and the requirements from the industry, together with architecture design and tool support in the Eppinger-Salminen framework it is possible to draw Figure 1-7 showing the scope of this thesis. The figure shows the focal areas of the thesis and points out that they all must comply with the requirements of cost-efficiency and dependability as an external concern. The dashed lines indicate that the relations between the different areas of special concern also are investigated within this work.

This thesis treats the issues of engineering an embedded system in an automotive context. An extended background on the technologies and applications of automotive embedded systems, the product aspect, is provided in chapter 2. The requirements of the industry concerning cost and dependability are further developed in chapter 3.

Chapter 4 discusses and relates two approaches to deal with the growing complexity to maintain a manageable lifecycle process; either to apply process measures such as formlized data management or to organize the product itself, decomposing it into an analyzable system architecture. The two approaches are efficient in different contexts, but also influence each other. The chosen process also influences the freedom in architecture design.

The relationships are indicated by the dashed lines in Figure 1-7. The chapter also discusses general aspects of models and model based development supported by computer tools. Proper data management, in this thesis referred to as model based development (MBD), can be utilized for architecture design but also for other purposes, for example to maintain analyzability.

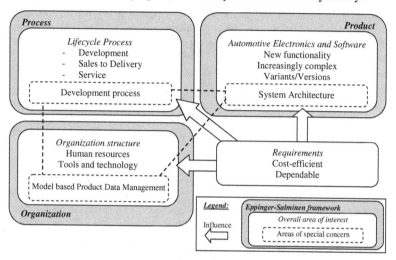

Figure 1-7 Thesis Topics and Scope

Designing an architecture is to define the structure of major building blocks in a system. Finding a sustainable systems architecture requires a considerate development process as illustrated by the dashed line in Figure 1-7. The architecture design process is covered in more detail in chapter 5. The chapter proposes a model based method for architecture synthesis and analysis. A range of aspects that must be considered in the architecture design process are also discussed.

In chapter 6 of this thesis the possibility to utilize model based approaches for systems engineering are explored. Information models for automotive embedded systems are proposed and evaluated in case studies.

The final chapter summarizes some of the main conclusions and also provides some pointers for further work in the field. The remainder of this chapter provides details on the CODEX project, the background, motivation and purpose of the project itself are presented.

1.2.3 Theoretical sources and related domains

The theoretical influences were drawn from a broad field in response to the broad topic of the research. Much of the early work aimed at grasping the basic theories of several academic communities in order to see how they applied to the research questions posed. The theoretical sources were found through courses, searching internet databases and references from colleagues. The final theoretic foundation has been collected from a wide variety of academic papers, books and other publications as well as input from attending relevant conferences from several academic communities.

For knowledge on the more practical issues of automotive electronics the society of automotive engineers (SAE) has provided a plentiful source of references. Specifically the SAE Convergence conference has provided helpful material. Also the umbrella organization International Federation of Automotive Engineering Societies (FISITA), of which SAE is a part, has broadened the views on automotive electronics in a global perspective.

The topic of architecture design is at the heart of the systems engineering community and the International Council of Systems Engineering (INCOSE). The related conference together with references from the Systems Engineering Journal managed by INCOSE has given insight into this field of research.

An even broader picture on the engineering design process is provided by the research performed within the Engineering Design community where references can be found to several frameworks, engineering methods and processes. The related conferences International Conference on Engineering Design (ICED) and the Design Conference gather research on general principles of problem solving as well as more detailed specific engineering solutions, design methods as well as supporting tools.

For Computer engineering, Software engineering and Embedded system design IEEE has provided valuable input through their extensive document database. Exchange within the Swedish real-time systems association, SNART, and the Swedish graduate school in embedded real-time systems, ARTES have given opportunities to gather a background in the field. International connections with the community for model driven engineering related to the development of UML have also provided some pointers to the different aspects of modeling in general and UML in particular.

Through INCOSE a broader view on modeling has been collected with references to SysML and the STEP standardization effort, STEP has also been partially explored in relation to the topic of product data management (PDM) with hands on experiences and local collaboration with the department of Industrial Production within the school of Industrial Engineering and Management at KTH.

In Sweden, Mekatronikmöte and several seminars with industry and academia with research in Mechatronics have provided valuable pointers to perspectives on research and resources in the field.

1.2.4 The quality of research findings

As mentioned, the major source of data for this thesis is literature studies and case studies. Another large source of data is observations made while working at Scania CV AB. Other sources of data have been conferences, meetings and social events where discussions and presentations have given an extended background of opinions and areas of concern for different stakeholders. Influences have been collected in a variety of communities, in industry and academia, at suppliers and manufacturers, in Sweden and abroad. This together with the combination of sources including experiences from actual work, conversations, participating observations and literature studies improves the internal validity, i.e. the credibility of the results.

Considering the external validity of the observations at Scania, i.e. the possibility to generalize the results to the entire automotive sector and beyond, the risk of being colored by a few groups at Scania is inevitable. Also, the corporate culture influences the ideas put forth. However, as the project assignments have been at four different groups and with a variety of case studies, the external validity is improved by multiple observations. Further, the external validity is also improved by inputs from other companies at conferences and similar events. Still, the results of this thesis must be interpreted from a perspective of emphasis on heavy vehicles in general and Scania in particular.

The reliability, i.e. the repeatability of the research is strong for the parts of the case studies based on implementation of tools. The theoretical and analytical parts of the case studies contain more qualitative evaluations and reasoning that may be more difficult to repeat. However, as summarized by Golafshani [2003], since reliability issue concerns measurements then it has no relevance in qualitative research, reliability is a consequence of the validity in a study.

1.3 Research questions and approach

This thesis has been written as part of the CODEX (COst efficient and DEpendable X-by wire systems) project, a collaborative effort between Scania and KTH. Initiated in December 2001 and currently intended for finalization at the end of 2005.

1.3.1 CODEX project purpose and goals

The purpose of the research in the CODEX project is to find a dependable and cost-effective x-by-wire system architecture for safety critical applications in heavy trucks. The work is motivated by the increased reliance on electronics for automotive applications in combination with the industry specific requirements for cost and dependability [Törngren et al 2001].

The early parts of the project have provided insights that have led to a shift in the objectives and project contents in response to these findings. The focus has broadened from the product architecture itself to also incorporate methods of system design. The purpose of this thesis is to provide methods and preconditions to design a cost-efficient and dependable automotive embedded system.

The project has maintained a direction towards the initial goal of improving the cost-efficiency and dependability of x-by-wire systems, but the hypothesis of the means to reach this goal has shifted. In the early phases of the project a general understanding of available by-wire systems were in focus, the first thoughts were addressing the reliability of the hardware in the system and the required level of redundancy for automotive systems. With the completion of the licentiate thesis [Larses 2003c] the problems and direction of research mainly concerned improving safety and maintainability, both through the design of the product architecture and through the actual development process.

The Licentiate thesis

In the licentiate thesis the long term objective of the CODEX project was stated as: *"to find guidelines to make well-informed decisions between alternative **solutions** and design **methods**"*, indicating both a product and a process perspective.

For the licentiate thesis six research questions were initially posed:

1. What by-wire systems exist today and what techniques and technologies are utilized?

2. What are the architectural options? By what methods are they designed?

3. What are the implications on dependability and cost of a specific solution or solution method?

4. How can guided choices be made between different alternatives (of solutions and methods)?

5. What is desired for a future truck drive-by-wire system? What system functionality needs to be supported?

6. How should this functionality be implemented in respect to cost and dependability? What alternatives are available?

Of these questions the final two concern specific solutions where market aspects must be considered. Avoiding such considerations, these business-related questions were practically left out of the licentiate thesis and they have not been covered in later work. However, a rapid growth in functionality is generally assumed.

In the licentiate thesis it was found that for future success, the embedded control systems carrying the by-wire functionality must be based on architectures that handle integration with external systems as well as integration of vehicle internal systems. The external integration should be handled through standardized interfaces. The internal integration also benefits from standards but in addition faces more stringent requirements on cost and dependability that must be managed by the OEM.

It was further concluded that the dependability concerns of internal integration should focus on safety, reliability and maintainability. In order to supply cost-efficient solutions, methods such as layering of the system and modular solutions are inevitable. Modeling is a supporting technique that should improve dependability, but the modeling should be accompanied by an extended reuse of designs to be cost-efficient.

Modeling can be applied formally or tacitly. In order to support modeling and development of either type a conceptual model developed in the licentiate thesis can be used. This model, labeled the Monty model, underlines the types of models necessary at different development stages and suggests a way to separate development stages in the development process. It is proposed that it is necessary to have models of the function, the implementation and the environment of the system. The model is thoroughly treated in a report by Larses & Chen [2003].

The results from the licentiate thesis show that technology is not limiting the feasibility of by-wire systems. Promoting cost-efficiency and dependability in the development process provides a larger challenge, including process and organization issues. Within dependability, methods to improve reliability are rather well known, implying that safety and maintainability requires more support.

Research questions for Doctoral Thesis

At the point in time when the licentiate thesis was finished the understanding of the problem to achieve a dependable and cost-efficient embedded control system had shifted the objective of the project from the product architecture to also include design considerations with respect to the complete lifecycle process. The long term objective of the licentiate thesis still holds: to find guidelines to make well-informed decisions between alternative solutions and design methods. It is however important to recognize that by choosing methods the architectural options are reduced, and also that a given architecture must be accompanied by a proper process. This thesis discusses these relationships and also proposes a set of design methods that are expected to be appropriate for the requirements of automotive electronics. For the doctoral thesis a new set of research questions is posed, renewing and slightly modifying the licentiate thesis questions:

1. What properties of a system improve cost-efficiency and dependability?

2. What methods can be utilized to ensure these properties?

3. How is the product architecture related to methods used in the lifecycle process?

4. How can guided strategic choices be made between different alternatives (of solutions and methods)?

5. How should tool support be utilized?

The performed research covers how different product features together with design methods, such as model based development and utilization of product data management (PDM) systems, influence dependability and cost. Methods to perform design, management and documentation of the product architecture are evaluated and proposed.

1.3.2 Project activities and sources

The conclusions in this thesis are drawn from a wide variety of influences. The research up to the licentiate thesis had a strong emphasis on literature studies. A rotation program at Scania included in the first year of the project also provided valuable insight into the problems of embedded system development in the automotive sector. The post-licentiate research has been predominantly empirical and driven by case studies. Working with the case studies several tools have been developed in order to support the work and evaluate theories. The tools have been used as means to evaluate the theories and were developed to prototype maturity, they were not intended to be further developed and evaluated for further public use.

The case studies at Scania have been performed in a setting where the results have been immediately applied and evaluated by the involved engineers. In addition to the work at Scania, a case study on the development of a model truck with an embedded control system has been performed as a larger student project at the Mechatronics lab at KTH. Besides the case studies the results and conclusions in this thesis have been derived from literature studies, discussions at conventions and conferences and also from personal experiences from the author working in product development projects at Scania.

Scania engineering project work

Participation in a number of projects has provided good experiences and research material. During 2002 and partially during 2003, engineering work was performed at four different organizational units (at that time named NTE, NME, RTSA, and RTSB) within Scania CV AB. Each posting lasted for approximately three months. The tasks included CAN-communication analysis, architecture analysis, signal analysis, function behavior analysis, function debugging and C-code implementation among other things. The rotation programme has given knowledge both through experiences with own work, and also through discussions and participation in meetings where issues of embedded control have been discussed.

Other involvement in projects at Scania has been related to the research project and has concerned further architecture design work and information modeling. This involvement has provided immediate feedback and practical perspectives on ideas and proposed methods.

Research studies

In the licentiate thesis it is stated that: "*In the future work less emphasis will be placed on literature and a stronger focus will be placed on case studies in building evidence for developed theories.*" This has been realized in the latter part of the project covering both methods for architecture design as well as model based approaches of embedded system design.

The architecture research has been performed in relation to ongoing work with architecture analysis and design at Scania CV AB. Ideas that try to shift architecting from an art to engineering have been tested in a live setting and experiences drawn from the results have provided material for the academic work. In total three different architecture case studies at Scania have been performed (case #1-3), each providing material for several different partial studies applying different analysis and synthesis methods.

Table 1-1 Academic studies in the CODEX project

#	Topic	Reference	Purpose	Scope	Related Tools
1	Architecture design methods	ACC	Evaluate method for architecture *synthesis*	Design of a single function (adaptive cruise control)	MDSM
2	Architecture design methods	PTA	Evaluate method for architecture *analysis*	Design of a sub-network	KFM, AIDA2
3	Architecture design methods	CTA	Evaluate *combination* of architecture *analysis* and *synthesis* methods	Design of a complete truck control system architecture	KFVB, MDSM, VDSM
4	Model based development	SAINT	Evaluate mechanisms and process for a configurable system platform	Evaluation of model and process support for a truck system	MVC
5	Model based development	FDoc	Function documentation supported by models	Modeling of functions in embedded systems	-na-
6	Model based development	Monty	Establish general system principles through Meta-modeling	A meta framework for General systems	-na-
7	Model based development	MBD Drivers	Establish drivers for MBD	Swedish automotive industry	-na-
8	Model based development	MBD Maturity	Establish an explanation model to evaluate the Maturity of MBD	Swedish automotive industry	-na-

The modeling research has been performed by a student project at KTH (#4) and a case study at Scania (#5), including practical tool implementations. Besides the tool based case studies, three more theoretical studies have been performed building on other research and experiences. One of these was performed in the work with the licentiate thesis and concerned the development of a meta-model for systems engineering, labeled the Monty model, which was exemplified through automotive examples (#6). The second built on studies performed in the Swedish automotive industry by Adamsson [2003] and aimed to identify drivers for model based development (#7). The third was a discussion on the maturity of model based development in different engineering fields related to mechatronic systems (#8).

All the research studies including references to tool implementations are listed in Table 1-1. The actual research approach and performed studies are in line with extensions proposed in the licentiate thesis, where it is mentioned that further research can be performed targeting *support tools, models and languages*; *function models*; *mapping models* and *the design process*. These topics are covered by the different studies in the table.

1.3.3 Case study backgrounds

To provide a better overview of the case studies a brief background on each of them is provided here. A more thorough treatment of the different case studies is referenced within the brief overviews, and further details on results and conclusions are distributed throughout the thesis.

Adaptive Cruise Control (ACC)

The purpose of the ACC study was to evaluate a synthesis method for a modular system architecture based on the design structure matrix (DSM). The case is discussed in [Larses & Blackenfelt 2003] and also in a Scania internal report.

Adaptive cruise control (ACC) may be seen as an extension to the conventional cruise control, where ACC not only keeps the speed but also ensures a given distance to the vehicles ahead, by using the brakes. The ACC is mainly seen as a comfort oriented function, although it could be seen as the first step towards a more autonomous driving. In the future this step could be followed by various functions aimed at comfort, safety and fuel economy.

In this study the new function was mapped to the existing Scania architecture, but the architecture of the function could be freely designed within the given constraints. These constraints were modeled together with the product elements and the modularization purposes. To keep down the complexity of the study, only the ACC function of the truck was considered. For the implementation of this function both new hardware and software was needed; new and existing elements needed to be grouped to modules.

For the ACC, some device to measure inter-vehicle speed and distance, typically a radar unit, is needed. These devices often include an ECU powerful enough for both signal processing and ACC controllers. Software for the longitudinal control may thus be placed in this unit or in any other ECU of the network. The function needs to utilize some ECUs that already exist in a modern truck. ECUs such as the brake management system (BMS), the engine management system (EMS) and the gearbox management system (GMS) provide necessary services. Also, a yaw rate sensor is needed and is available in the BMS but a dedicated sensor could be introduced. Further, a vehicle speed signal is needed that may be received from various available sensors. The role of the existing and new hardware needed to be defined and the new software needed to be mapped onto the hardware. The goal was to provide a solution as modular as possible, for this purpose cluster analysis utilizing the DSM was performed.

Partial Truck Architecture (PTA)

The purpose of the study was to evaluate methods to analyze system architectures by keyfigures. The case is described in Scania internal documents and some of the lessons learned are discussed in [Larses & El-khoury 2004].

The architecture design problem was approached in a project of designing a partial EE network architecture at Scania, here referred to as the PTA project. The project aimed at developing an improved solution for a bounded part of the EE architecture. It was assumed that the partial architecture would be clearly analyzable on its own.

Based on the intended functional content of the partial architecture an allocation of functions to components and a geometrical distribution of components was used as input data for the utilized analysis tool. The goal was to perform a trade-off to minimize the cost of the implementation over the product range. For this purpose a low end vehicle, a typical standard truck, and a high end vehicle were evaluated in the tool. The tool was able to calculate keyfigures for properties of the resulting hardware and allowed a minimization of cabling and hardware components across the three truck variants.

Complete Truck Architecture (CTA)

The purpose of the CTA study was to evaluate the combined value of the analysis and synthesis methods for automotive embedded system architectures, developed in the ACC and PTA case studies. The combined method is documented in a Scania internal document and also in [Larses 2005a].

In the case study the complete truck architecture of Scania was approached, product data for the complete functional content and component set of the Scania product was collected in an Access database and the data was manipulated by a tool developed in Visual Basic. The experiences from the tool developed in the PTA study were used as a basis for the development of the new tool. The algorithms and scripts implemented for the DSM cluster analysis in the ACC case were also reused in a separate parallel effort. The combined toolset provided support both for analysis and synthesis in the architecture design process. Further details are mainly provided in chapter 5.

The case study progressed by an iterative process of tool development, using the tools for their purpose and evaluating the result in the engineering group at Scania.

Function documentation supported by models (FDoc)

In the FDoc case study the documentation and roles regarding function documentation at Scania were studied with an information modeling perspective. The work was delimited to concepts and product data regarding the EE (Electrical-electronic) system with a special focus on the concept of function. Using an information model it was possible to analyze the current documentation solution and suggest improvements.

The study was performed by Ola Larses and Jad El-khoury from May to October 2005. The case study was performed in three stages. First an initial pre-study was performed where the current situation at Scania was investigated and a hypothesis for the modeling of functions was developed. The second stage concerned presentation and discussion around the proposal to verify the hypothesis. In the final stage some practical work was performed to try the proposed modeling approach in practice.

Throughout the study, three different user functions have been used as case study material: *Adaptive cruise control, parking brake warning & cruise control*. Each of the case studies has served different purposes.

Adaptive cruise control was the first case, used as a rather advanced function to explore different concepts and ideas. Different representations and general modeling guidelines were investigated through this case.

The second case study, parking brake warning, provided an example where the ideas could be applied to improve the current document structure. In the process, a complete set of documentation of the studied user function was developed based on the new proposed document structure. The work resulted in a template document as well as an illustrative example document.

The cruise control was the third case study. This acted as a test of the improvements of the documentation suggested in the second case study. The study is further discussed in chapter 6.

Self Adaptive Intelligent Truck (SAINT)

The SAINT project was a student-project at KTH that for almost one full semester employed sixteen final year students pursuing a master degree in mechatronics. The purpose of this case study was to evaluate a range of measures to improve the modularity and configurability of an automotive system. The design of the product architecture was combined with tool supported process measures to provide a broad variety of configuration options. The project is documented in a Scania internal document and in a student report [Blixt et al 2005].

The project was sponsored by Scania Trucks and included the task to develop a configurable demonstrator. The demonstrator was built as a wirelessly controlled scale 1:6 model truck and trailer, with an electrical motor, brakes and a distributed and embedded control system. The system solution, with the truck and trailer together with the external components for truck operation and configuration, is shown in Figure 1-8. Support for advanced and configurable functionality such as collision avoidance, steering on rear wheels of the trailer and load sensitive alarm was implemented. The focus in the project was however not the advanced functions, but rather developing and testing a methodology for flexible function development and configuration, supporting the three processes mentioned in section 1.1.1. For this purpose a number of applications to manage software configurations were built around the product data management (PDM) tool Matrix 10. PDM tools are generally not applied for this purpose, software is usually managed in software configuration management (SCM) tools, and the case study provided a possibility to examine how a PDM system can be used to support a process for embedded control systems. The architecture of the demonstrator is described in chapter 2, while model issues are discussed in chapter 6.

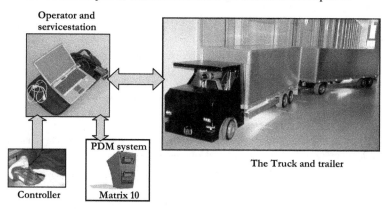

Figure 1-8 An overview of the SAINT demonstrator system

1.3.4 Theoretical studies

Similar to the more practical case studies a brief background on the theoretical studies is provided here.

Monty

The purpose of this case study was to provide meta-level descriptions of important properties and relations in both products and design. The idea was to form an objective basis for assessing and linking modeling and analysis techniques, reasoning about design traceability, and comparing various process models and development concepts. The study is thoroughly descried in Larses & Chen [2003].

The study was based on a comprehensive literature study on the topic of system design and provided a meta-model referred to as The Monty model. The resulting model, elaborated in chapter 6, comprises of two sub-models.

- A general meta-model of systems referred to as the Monty Meta-Model of Systems. The model is indented to be an easily understandable framework for identifying and relating various system aspects.

- A generic model of system design that covers fundamental issues of engineering, solutions, and their links, referred to as the Monty Model of System Design. The aim is to provide more precise characterizations of design dynamics, supporting the reasoning of some fundamental aspects such as traceability.

MBD Drivers

The purpose of the MBD Drivers case study was to establish drivers for model based development (MBD) using theories of knowledge management. The case study is presented in [Larses & Adamsson 2004].

The basis for the discussion in the paper is results from an empirical case study of MBD performed in the vehicle industry in Sweden performed by Adamsson. The case study was based on interviews at three different vehicle-producing companies [Adamsson 2003]. An explanation model for the empirical findings is derived from earlier knowledge management theories, and the model is used for an explanation of the results from the study.

MBD Maturity

The purpose of the MBD Maturity case study was to provide an explanation model for observed objections and opinions about MBD by comparing the maturity of model based development processes and tools in different engineering disciplines with the drivers for MBD. The case study is presented in [Törngren & Larses 2004].

The study draws upon the study in the Swedish automotive industry by Adamsson [2003], but also on other published references as well as personal experiences of the authors including informal communication with industrial players. In the study a model that enables comparison of the maturity of model based processes, the maturity of available tool support and the drivers for model based development are introduced and used for analysis of the automotive sector.

1.3.5 Developed Tools

For the case studies a number of tools were developed. Two tools supported cluster analysis. The first of these tools (MDSM) utilizes Matlab to automate cluster analysis. This tool has a very simple user interface reached from the command line of Matlab. In the CTA-case a tool consisting of a complementary Excel-sheet with a set of macros (VDSM) was utilized to manage the DSM data.

Three different tools target keyfigure analysis. In the first tool (KFM), analysis was first performed in an Excel-sheet but then required migration to a Matlab environment to cope with some calculations too cumbersome for Excel to handle. In parallel, another tool (AIDA2) was developed in the Dome environment by Jad El-khoury [2005]. The third tool (KF-VB) built on the experiences from the first two tools and was developed in collaboration with El-khoury. A visual basic application that used an Access database improved the usability and performance of the tool.

A completely separate tool was implemented to support the product data management of the SAINT-project. The MVC-tool mimics the very basic core functionality of the CVS version management tool but uses the MatrixOne PDM-system for management and storage of files.

The different tools are summarized in Table 1-2 and further discussed in relation to the topics they cover. It should be noted that the SAINT project also incorporated the development of a configuration environment connected to the PDM-system. This implementation is not included here but is seen as a part of the project.

Table 1-2 Tools implemented in the CODEX project

Tool reference	Used in Case	Implementation environment	Purpose	Topic
KFM	PTA	Excel/Matlab	Keyfigure analysis	Architecture design
AIDA2	PTA	Dome [El-khoury 2005]	Keyfigure analysis	Architecture design
KFVB	CTA	Visual Basic/ Access	Keyfigure analysis	Architecture design
MDSM	ACC, CTA	Matlab	Automated Cluster analysis of DSM	Architecture design
VDSM	CTA	Excel	Visualization of clusters in DSM	Architecture design
MVC	SAINT	MQL/Tcl/Matrix	Adding files to the matrix repository for version management	MBD

Keyfigure tool – KFM

To analyze and compare architecture designs a set of keyfigures were developed. The keyfigures provided a benchmark for different architecture alternatives. An Excel workbook called *PTAanalysis.xls* was developed for the analysis. The structure of the references across the sheets in the workbook is presented in Figure 1-9, each box in the figure represents one sheet in the workbook. In the sheets it was possible to enter data on connections, function blocks (analysis blocks in the figure) and product configuration (conditions and analysis in the figure). With provided data the spreadsheet calculated a range of keyfigures as indicated by the 10 boxes in the analysis parameters frame (*AKP, AK, AL,* etc.). The *Analysis blocks* are the core of the analysis tool with the relations between them provided in the *Relations* matrix. The *Connections* and *Interface* sheets provide data on the implementation of the relation. The calculations of specific parameters are provided in separate sheets, one per parameter. The *RI* sheet allows grouping of analysis blocks to change the architecture. *Conditions and Analysis* provides the product configuration that selects the analysis blocks and interfaces that should be included in the calculation. In the sheet a set of logical conditions are specified for this purpose. The sheet also summarizes the results of all calculations and thereby becomes the master sheet for the application.

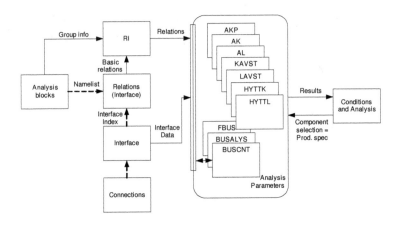

Figure 1-9 Relations among sheets in the Excel workbook of the KFM tool

The calculation of the final keyfigures from the data required a number of matrix operations. As the number of keyfigures grew Excel was not capable to deal with all the calculations and the raw data in the Excel sheets was transferred to a Matlab environment where calculations could be performed much more efficiently. The main parts of the basic infrastructure were kept but the way keyfigures were calculated was improved. The communication with the user was performed through a graphical interface implemented in Matlab. A screenshot of this interface is shown in Figure 1-10. In the interface the upper part provided a possibility to define a given product variant and the lower part provided a keyfigure analysis.

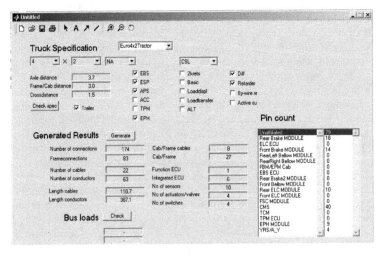

Figure 1-10 The Matlab interface of the KFM tool

Keyfigure tool – AIDA 2

The AIDA2 tool was developed by Jad El-khoury in the Dome environment and used in the PTA case study. This tool and model is described in further detail in [El-khoury 2005].

Keyfigure tool – KFVB

The keyfigure tool developed for the CTA case study was based on an Access database that held all relevant product data. For the keyfigure calculation, the database required collection of data from a range of dispersed sources. Many of the data sources were adapted by manual efforts and entered into the database. A core source of information was a function database that was converted to fit the purpose. Component data of the existing implementations was collected from a Component library. This data included the location of components that was converted to an enumerated position. Another important source of implementation data was the Scania internal CAN-specification. For configuration, to create descriptions of product individuals, data was collected in the form of Scania proprietary product identification files. Some of the data was available on machine interpretable format, such as databases, but some of the data was not machine interpretable at all, such as word-documents which made the collection more difficult.

Details of the functions were provided by importing message sequence charts (MSC) to the database from models available in the UML tool Tau from Telelogic. This data was converted to the database format by a tool implemented in Visual Basic. The architecture of the tool and the different sources of data are shown in Figure 1-11.

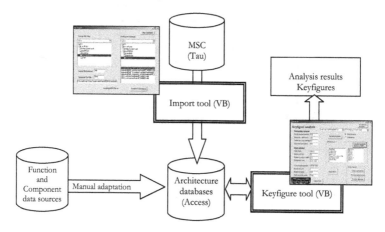

Figure 1-11 Tool architecture for the keyfigure calculation case study

In the database a clear separation between function data and implementation data was achieved. Functions were (in principle) modeled as function blocks linked by associated communication links. The implementation was (in principle) modeled as electronic units linked by associated cables. The electronic units include sensors, actuators and electronic control units (ECU). In the database, function blocks are associated to electronic units and communication links are associated to one or more cables. These associations show the implementation of functions and communication links as software and signals in the physical system. The UML sequence diagram data stored in the Telelogic Tau tool was exported to a text format and then automatically imported into the Access database as representations of function blocks and communication links. The remaining data conversion was performed manually through cut and paste efforts. Further information about the tool is provided in chapters 5 and 6.

Design Structure Matrix (DSM) cluster analysis tool – MDSM

The MDSM tool is a set of scripts developed for Matlab. The tool analyzes an architecture defined by a DSM matrix according to the method described in section 5.5. Provided with a matrix representing relations between objects, the tool combines the objects into clusters where the positive relations within the cluster are maximized and the relations between the clusters are minimized. The tool also delivers quantitative measures of the quality of the clusters, definitions of these measures and a more detailed description of the tool is provided in section 5.5.

Visualization of clustering – VDSM

To improve the management of the DSM analysis an Excel-sheet vas implemented with macros that interact with the MDSM tool. In the Excel sheet it is possible to balance weights between DSMs of different aspects and graphically visualize the resulting total DSM in a diagram, as shown in Figure 1-12. The Excel sheet provides an ability to export DSMs to the file format readable by the MDSM tool, and import clusterings and ARP data from the files produced by the MDSM tool. Further, macros are available to reorganize the DSM according to the clusters provided by the analysis.

Version control – MVC

To support the SAINT project a simple tool that could interact with the Matrix PDM system was developed. MVC is a simple version control tool that mimics some basic functionality of the batch version of the CVS (Concurrent Versions System) tool [CVS 2005]. MVC is implemented as an extension to the Matrix PDM (product data management) environment using the Matrix specific query language MQL and the script language TCL. The implemented tool supports the commands shown in Table 1-3. The tool uses modules to collect a number of files that are managed together in the system.

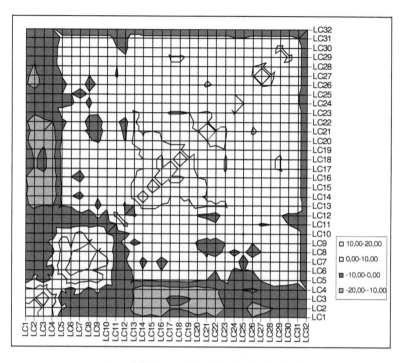

Figure 1-12 The graphics of the visualization tool

Table 1-3 The MVC tool commands

Command short	Command name	Description
add	add	Adds a file to version control under a given module
co	check out	Retrieve a file (or a set of files) under version control from the repository
ci	commit	Check in a file (or a set of files) under version control to the repository
diff	difference	Show difference (on ascii level) between two revisions of files
help	help	A help command is included to provide references
init	initialize	Create a new module to which files can be attached
log	log	Examine history of file in the repository
merge	merge	Commit files and override repository
modules	modules	List available modules in repository
remove	remove	Remove file from version control
update	update	Updates files in the local directory to comply with repository
verbose	verbose	Toggles the amount of output to screen

2 Background – Automotive Embedded Systems

This section provides a background on automotive embedded systems from a technology perspective. First some general points of embedded systems are discussed, and then automotive applications are introduced. One section gives special attention to drive-by-wire systems before the technologies and common architectures of automotive embedded systems are described.

The chapter is based on [Larses 2003a], and the introduction on embedded systems is based on [Törngren & Larses 2004].

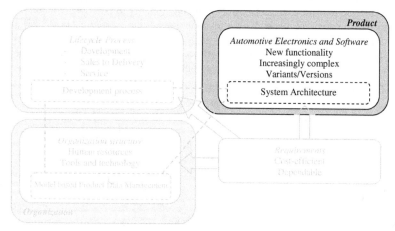

Figure 2-1 Chapter focus - Product

2.1 Embedded systems and Mechatronic products

Automotive systems are a specific application of mechatronic systems, that is, systems which are formed by an integration of mechanical, electronic, software and control system components. Mechatronic systems are traditionally associated with functionality involving controlled motion of mechanical devices. The electronics in the system is interacting with the physical environment through sensors and actuators. Non automotive examples of such systems include medical equipment, robotics and manufacturing equipment.

Software, electronics, sensors and actuators constitute key implementation technologies in mechatronic systems. For low-end products, analog technology can be, and is being, used. However, software is the dominating technology in more complex applications. The standard solution today is to use software with microcontrollers (highly integrated electronics devices that include a microprocessor, communication facilities, and digital/analog inputs and outputs). The use of electronics and software within products has given rise to the term embedded systems. IEEE [1992] defines an embedded computer system as:

A computer system that is part of a larger system and performs some of the requirements of that system; for example, a computer system used in an aircraft or rapid transit system.

Embedded control systems are increasingly used in a variety of applications and they are radically changing the products they are embedded into. The added dimension of explicit and flexible information transfer and processing, implemented through electronics and integrated into the mechanics, enables improved performance and entirely new functionality to be implemented. The use of embedded control systems has paved the way for advances of machinery. Simultaneously, the integrated and distributed functionality introduces new, and sometimes hidden, dependencies among components that must be dealt with in the development process.

To develop mechatronic products, integration of multidisciplinary competence in product development is required. New development processes and organizational structures must be established, matching the new product structures [Eppinger & Salminen 2001]. Trying to achieve synergistic effects in the integration is often referred to as a mechatronics approach in development [Wikander et al 2001]. Such an approach conforms to traditional systems engineering and emphasizes co-design and systems optimization.

Besides the multidisciplinary technology content, mechatronic embedded control systems are characterized by [Törngren & Larses 2004]:

Tight coupling to the environment. The control system is fundamentally related to the controlled process and the interaction with the environment. Understanding these dependencies is essential for dependable service. To facilitate design, models of the environment are commonly used in control design. The control algorithms may be synthesized from a validated model of the controlled system (model based control systems development). In other cases, the controller parameters are tuned based on the overall (closed-loop) system behavior.

One of the most difficult problems is humans interacting with the embedded control system. Many control systems for example have an operator "in the loop". This is typical for vehicular systems, and the situation arises where conflicts can occur – who is deciding the motion of the vehicle at any given point in time? Careful analysis is required and special care has to be given to the human/machine interface.

Real-time constraints. The control systems must be able to cope with the requirements and constraints derived from the controlled system. Based on required speeds of motion, precision and time durations, timing requirements on the embedded control system can be derived. For example, the speed (or bandwidth) of the closed loop system will provide requirements on the timing of the controller, including the frequency of sampling periods that can be used, and feedback delays and jitter that can be allowed. Some properties can be taken into account in the control design, however this introduces further dependencies between the control design and its implementation.

Diverse services. Motion control is the central part of embedded control systems, however additional services are commonly added. The systems are often structured into a system platform and applications. Services of the system platform include for example initialization, logging, basic communication services, drivers for sensor readings and outputs, and error detection. Besides motion control, applications may include diagnostics related to the control, estimation of the environment state, and human machine communication. The different applications and services may have different requirements or compete for the same resources introducing more concerns for the system design.

Distributed systems. There has over the last decades been a strong trend to connect stand-alone controllers by networks, forming distributed systems. Distributed control systems first appeared in process control in the 70s. It was utilized for the first time in aerospace in the 80s and in the automotive industry in the 90s.

The dispersed locations of sensors and actuators makes the hardware distributed. In addition, with several control units different distributions of data and of control (decisions) in the system can be considered. Distributed systems are characterized by the mapping problem, i.e. the need to assign functions to different nodes of a distributed system and their implementation in software and/or hardware.

Related to distribution is the forming of integrated mechatronic modules, where an electronic control unit is physically integrated into a mechanical component such as an engine. Combining the concepts of networks and mechatronic modules makes it possible to reduce both the cabling and the number of connectors, resulting in easier production and increased reliability of the system. At the same time, the software systems grow and hidden interrelationships are introduced increasing the complexity of the embedded systems. Managing the introduced product complexity is becoming a crucial problem and a challenge for the automotive industry today.

Sampling and triggering. A typical control system normally includes both time- and event-triggered activities. In many cases, time-triggering follows naturally from the development of discrete time (sampled data) control functions. However, in other cases the controlled process can be inherently event-triggered. For example, this is the case for the control of injection in a combustion engine; the point in time of injection depends on the speed and angular position of the engine parts [Åström and Wittenmark 1990]. Coping with the combination of event-triggered and time-triggered functions can be a challenge.

Resource constrained implementations. Embedded control systems are often highly resource constrained. This is applicable also for automotive applications where the large series produced provide strict cost constraints on hardware, a small cost increase for one unit multiplies to a huge total cost. In such applications, trade-offs between the system behavior (quality of service) and the resources required (processing, memory and power) is essential.

2.1.1 Complexity

Another commonly referred property of automotive embedded control systems is the increasing complexity. One fundamental issue in embedded system development is the need of managing complexity. The consequences of unhandled complexity resulting in design errors can be severe, introducing increased development cost, delayed product delivery, and even major catastrophes if the system is safety-critical.

In general, complexity is a relative concept. It is concerned with the degree to which a system can be discerned, communicated, and verified by the stakeholders of a system. This in term depends on the levels of details upon which a system is being considered, the ways by which the information of the internal and external aspects of a system is structured and represented for the stakeholders, and the knowledge or mental models of stakeholders about the system [Leveson 2000].

While complexity is mainly a subjective attribute, several inherent factors of systems make complexity management a challenging task in system developments. Three important factors that contribute to complexity are discussed below, including *conflicting requirements*, *richness of system content*, and *heterogeneity of rationale*.

Conflicting requirements

Most systems need to satisfy multiple requirements such as functionality, performance, and cost. Since these requirements can be interdependent, optimizing the system solution for a maximum total benefit is a difficult issue. Trading, comparing and modifying requirements may be necessary for a feasible solution.

Some of the requirements can have logical conflicts. For example, in a respirator, there may be a conflict in the sense that deviating operation may suggest a shutdown, but a shutdown may be equally hazardous for the patient. Other requirements can have dependencies and conflicts due to the underlying solutions such as a certain implementation technology. For example, a performance requirement may contradict the requirements of weight and power consumption. While the conflicts of requirements in logics are more obvious, their dependencies and conflicts due to underlying solutions may appear only at a certain refinement stage of design. If such conflicts are detected late in development, costly redesign is unavoidable.

Richness of system content

Inside a system, there can be a large number of constituents of various types and with varying properties and relations. The richness of system contents adds to the complexity by increasing our mental loads. Moreover, it can also be more difficult to ensure the consistency as well as to integrate these solutions due to the existence of multiple compatibility criteria.

This is for example the case of embedded computer control systems. There are application software programs for motion control and mode logic, system software for multitasking and resource management, hardware components for computation and storage, and electrical devices for communication with the external conditions. These constituents have not only operational relations such as communication, synchronization, sharing resources, allocation and execution, but also analytical relations established by the functions they attempt to implement such as performance, errors and reliability.

Heterogeneity of rationale

In system development, each design decision can have some implications on other design decisions. It is important that these implications are motivated and clearly documented. The documentation should include derived requirements or constraints in a way that they cannot be ignored as the design goes on. In this respect, the *heterogeneity of rationale* behind design decisions adds to the system complexity. That is, the transformations of information across different types of rationales can be difficult due to the absence of a common notion. This is often the case in the development of embedded computer control systems where multiple engineering domains need to collaborate.

The issue of constructing a homogeneous mental model of the system in order to improve collaboration and synergies on an organizational level has been discussed by Senge [1990]. It is identified that a set of common mental models is important for organizational learning, which is the target of research and development organizations. A homogeneous rationale improves system thinking and reduces complexity.

2.2 Automotive applications

Electronics is utilized for a wide range of applications in the automotive sector. This section introduces a framework for classification of applications and also discusses some typical automotive application classes. An application can be classified according to target and timing, further described below.

2.2.1 Target of applications

The target of applications tells you the role of the user that the application aims for. Applications can aim for the people in the car (personal applications) or the actual car (vehicle-centric applications). There are also some applications specifically aiming at the driver, combining vehicle-centric and personal applications. The framework is illustrated in Figure 2-2. The three derived application sets are labeled under-the-hood, front-seat and back-seat illustrating their target role in the vehicle.

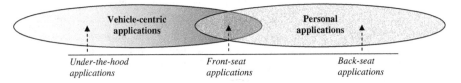

Figure 2-2 Target for applications

Front-seat

Front seat applications aim at the driving experience. The focus is on the driver in the role of a driver, not in the role of an individual. The applications can include passive support for the driver through information systems, where an obvious example is a navigation system. Active support is also possible, automating driver decisions. Some of the active support relieves the driver of mundane tasks like switching the windscreen wipers on and off in case of rain. Other active support functions can take over actual driving tasks like automated gear shifting and cruise control. Some of the applications are vehicle-centric, as the cruise control, while others are personal applications, like the navigation systems, but they all aim at the driver in the role of a driver.

Back-seat

These applications are aiming for comfort activities that are not related to the actual transport objective. Back seat applications are there for the people riding in the car in the role of individuals. This mainly aims at passengers but also aim at the driver as an individual. Obviously, the driver might also like to listen to music while he is in the vehicle. Some back-seat applications relate to communication needs and include use of mobile phones and mobile internet. The communication can be used both for leisure and for work issues introducing a truly mobile office. Other back-seat applications relates to the use of the vehicle for non-transportation functions, such as comfort facilities for the truck-driver that lives and sleeps in his vehicle while he is on a work schedule.

The back-seat applications, driven by the fact that we are spending more and more time in vehicles, bring consumer electronics into the vehicle [D'Avello & Van Bosch 2002]. Multimedia-entertainment is considered to be a major field for automotive electronics in the future [Schumacher et al 2002]. For this field of applications availability of services is of utmost importance, entertainment should be possible to buy anywhere, at any point in time.

The differences between front-seat and back-seat applications are acknowledged by the iDrive concept from BMW [Spreng 2002]. In order to avoid driver overload the intention of the iDrive concept is to use the distinction between driving and comfort applications and separate the controls for the two purposes in the drivers' compartment [Fuchs et al 2002].

Under-the-hood

Under-the-hood applications can either improve the handling of the vehicle or influence issues like maintenance, economy or the environmental impact. The applications are related to control and diagnostics of the vehicle. The vehicle control related applications are often referred to as drive-by-wire where electronics are used to collect and interpret the signals of the driver in order to create an actuator response as close as possible to the anticipated driver desires. Proper control of a vehicle can improve the economy in terms of fuel-consumption and wear of components. Electronic diagnostics can be used both during maintenance but also as a warning system if bad readings from sensors indicate mechanical, environmental or other physical hazards. The possibility to collect driving data for statistics is another obvious application.

2.2.2 Timing of applications

Modern vehicles also have applications that are utilized outside the time span of actual transportation. It is possible to distinguish between three categories of usage modes. The user can be riding, resting or parked as described below. It is also possible to consider a fourth category that can be labeled *workshop* and includes maintenance work that requires physical control over the vehicle, such applications are not further developed here.

Riding

Riding is the obvious standard situation while using the vehicle for transportation. This is the traditional focus for applications and also the kind of applications that always can be utilized by the user.

Resting

This category indicates that the user is in the vehicle but is not using it for transportation. Possible uses include a short stop for resting by the road, sitting in on a drive-in movie or making an overnight sleep in a truck.

Parked

Applications in the parked category relates to the time when the user is not in the vehicle. These functions need to be autonomous or remotely controlled and are to some extent an unexplored issue. Current applications include alarms and anti-theft systems. Important enabling technologies for these applications include external communication with the vehicle, an area that is expanding but has not yet become a standard issue in current products.

2.2.3 Classes of applications

Using the concept of application target it is possible to further refine classes of applications within the framework. In this section eight distinct classes are defined and discussed: Diagnostics, Environmental/Economy, External coordination, Safety, Handling, Driver support/ Navigation, Communication and Infotainment. The classes are based on different purposes or goals of the applications. The different goals are usually related to a specific stakeholder or role and it is therefore possible to arrange the application classes along the application target range. The classes range from applications with a vehicle centric focus to pure personal applications as shown in Figure 2-3. The application classes are further elaborated below.

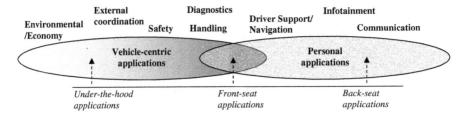

Figure 2-3 Classes of applications

Some of these application classes have overlapping properties and some applications may fit into more than one class. The framework introduced in this section is not intended as a strict classification tool but instead tries to place some common classification concepts in the more general framework introduced in previous sections.

Environmental/Economy

Environmental applications have the goal to reduce the impact on the environment from the vehicle. In parallel to this is the goal of improved lifetime economy of the vehicle. A lean process that consumes few resources is not only environmental friendly but also cheap and cost-efficient when it is running. Environmental issues are heavily driven by legislations and regulations. This field of applications includes the use of electronics to reduce emissions, change production to more environmental friendly processes, remove hostile substances and improve energy efficiency. The applications are vehicle-centric under-the-hood applications and are necessary efforts due to the strict regulation of the area. The economy aspect is more important for commercial vehicles where fuel-consumption and maintenance are important parameters.

External coordination

External coordination applications links vehicle external objects to the vehicle and allow them to influence the performance and/or behavior of vehicle internal functions. A commonly referenced application would be convoying. In convoying one vehicle is designated the role of convoy leader and other vehicles are allowed to hook on to the convoy as followers. The operation of the first vehicle is communicated back through the rest of the convoy. This allows immediate response to, for example, braking action throughout the entire convoy. With this coordination the vehicles can be driven with a very small distance between them which improves fuel consumption and also reduces road congestion.

Other applications could be exemplified by externally enforced speed limitation and temporary performance enhancements of the vehicle through software tuning. Within this application class few, if any, applications exist today. There are many unresolved organizational issues to deal with before vehicle systems can be fully integrated in higher level traffic management systems that coordinate and optimizes traffic flow.

Safety

Safety applications have the goal to ensure that no harm is done to humans in or around the vehicle. The safety applications are built into the vehicle, so even if safety applications target individuals it is a set of vehicle centric applications. The applications are under-the-hood or front-seat applications as they are usually deployed by the vehicle itself.

Safety applications are sometimes classified as active or passive applications depending on if they prevent accidents or reduce the impact of accidents. Some safety related applications also work to improve the situation after the accident like mayday signaling systems where a call-center is notified that the airbags of the vehicle have been deployed [Butler 2002].

Diagnostics

The purpose of diagnostic services have developed and diversified over time. Diagnostics originated as support for the workshop technician to improve efficiency in repairs. Diagnostics can also be utilized for monitoring of deterioration of components, instructing the driver to perform maintenance to avoid breakdowns. Further, diagnostic services can include legal compliance, environmental protection and diagnostic checks performed by dealers. The diagnostic services can be applied locally on the vehicle but also while the vehicle is operating or through remote access using telematics solutions. The remote access can be utilized by external actors to prepare the service organization for upcoming repairs or supply data for further product development [Ogawa & Morozumi 2002].

Handling/Vehicle dynamics

Applications aiming at the handling and dynamics of the car are generally referred to as drive-by-wire systems. Some of these aim at improving the drivers' impression of the vehicle, others aim at improving vehicle dynamics like braking distance and stopping distance. By-wire control applications include systems like non-locking brakes (ABS), electronic stability programs (ESP) and electronic control of combustion in engines. In modern applications, networked collaboration between by-wire subsystems is seen as an area where vehicle control can be substantially improved. An example of improved stopping distance by networking is given by the '30 meter car', using sensors in tires, distance sensors, active suspension and brake-by-wire networked with good results [Rieth & Eberz 2002].

Navigation/Driver support

Navigation/Driver support applications have the goal to assist the driver with tasks associated with driving the vehicle. Navigation applications are facilitated by the introduction of the GPS-system. These applications are basically front-seat applications assisting the driver with information about routes and, in some systems, also information about the current traffic situation. Other navigation applications are linked to the driver as a part of a commercial logistics system. These systems instruct the driver with destinations and can also sometimes help out with navigation. More advanced applications within this type are fleet management systems that coordinate the navigation of an entire fleet of vehicles. Driver support applications can either supply information to the driver or filter information by screening processes and refinement to a better information quality. Systems providing information include lane departure warning systems that send out a signal if the driver is unintentionally deviating from the current lane.

This class of applications also includes automation applications that relieve the driver of decisions, such as automated windscreen wipers and systems for managing incoming telephone calls. There exists research on how to measure the cognitive pressure on the driver, the aim is to balance the cognitive load of the driver by employing different sets of applications at different times [Pompei et al 2002; Aragane & Tsuji 2002]. A reasonable amount of information, producing a balanced cognitive load, enables better decision-making from the driver [Pompei et al 2002].

Communication

Communication applications have been around for a while in the shape of cellular phones and computer networks. The goal of these applications is to get and send information between individuals, which classifies them as back-seat applications. These applications are generally supplied by telecom companies and are already well developed. Integrated standalone car phones or hands-free capabilities are not new services. In Japan it is a legal requirement that a hands-free system should be used if a cellular phone is used while driving, such legislation increases the need for integrated communication solutions.

Infotainment

The infotainment applications are back-seat applications that integrate consumer electronics in the automotive environment. Cars can include video screens, game consoles and audio systems, almost anything that you can find in your home. As infotainment in the home is becoming computer and Internet based, trends indicate that the same services can be desired in vehicles [Bock et al 2002]. Paying for entertainment like on-line music sales is projected to be one of the fastest growing consumer markets over the next few years. This development indicates a rapidly growing market for automotive infotainment applications as access and storage of information and media become important services [Schumacher et al 2002].

2.2.4 Requirements driven by applications

Why is it useful to make the above distinction in target and timing for technical aspects? The target and timing of applications will imply requirements on technology. Understanding these requirements is essential to identify technology needs for the future, making it possible to develop a strategy on where to put development effort, where to collaborate and where to outsource the responsibility of development. Therefore a model of applications is useful if relationships between application properties and technical needs can be derived.

Many new applications rely on the vehicle being connected to an information infrastructure like the Internet at any time. This requires a wireless communication infrastructure like GSM or different wireless networking standards. It is necessary to be connected at any point in time. The need for connectivity is strongly related to the timing of applications.

Increasing the number of applications in the resting and parking segments also increases the need for efficient power management. If electrical power is generated by the engine only the storage and infrastructure of electrical power will be a difficult task. This issue is not further investigated in this thesis.

Furthermore, some of the applications relieve the driver of responsibility or even assume control of the vehicle itself. If any system that influences the driving of the vehicle fails the consequences can be grave. These applications are called safety-critical and can be either direct within the vehicle or indirect, distracting the driver to make mistakes. The criticality is strongly related to vehicle-centric applications, further developed in the section on dependability requirements.

Applications can be placed in a graph as illustrated in Figure 2-4. The traditional core application of the OEM with vehicle-centric applications running while driving are, due to the development of electronics, now accompanied by a range of market opportunities. To diversify into personal services while the vehicle is parked seems to be far outside the automotive core business. However, where to draw the boundary and how to integrate the applications externally developed, must still be dealt with in the system architecture of a modern vehicle.

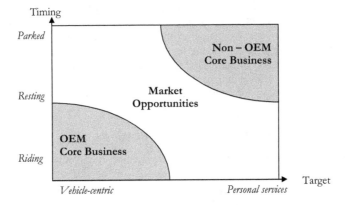

Figure 2-4 Market opportunities for automotive electronics

2.3 Drive-by-wire systems

This section describes the vehicle-centric applications known as drive-by-wire. The subject is discussed in the context of other by-wire systems. By-wire is a general term for the use of electronic control in any technical system. This section suggests a systematic approach to terminology and related concepts. Automotive applications are covered in more detail, including a discussion on associated benefits and also some problems that may need to be addressed before the introduction of full drive-by-wire systems will be possible.

2.3.1 Terminology and concepts

The terminology of by-wire systems has developed organically as new areas of application have been identified. By-wire concepts regard embedded control systems and this section suggests a structured approach to the terminology. The by-wire terminology can principally be applied on any embedded control system and in any area of application.

Here, a distinction between by-wire control systems and full by-wire systems are made. In a *by-wire control system* (BWC) some stage of the information transfer is performed through digital communication and digital decision making, not only analog signals and filtering is used. It is generally assumed that the information or control signals are digitally communicated by an embedded control system, consisting of a set of electronic control units (ECU) that are interconnected in a local network. In a *full by-wire system* (FBW) the information is digital and also the power is transferred electrically by the use of electrical motors and electromechanical actuators. Using the terminology it is possible to state that a full by-wire system is a by-wire control system with electromechanical actuation.

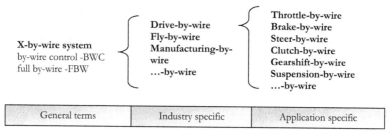

| General terms | Industry specific | Application specific |

Figure 2-5 By-wire terminology

A widely and rather informally used term is *x-by-wire*. X-by-wire is a general term that contains any application with by-wire functionality where 'x' is a referral to the actual application. Further concepts can be developed based on the application scope of the system. The resulting set can be arranged hierarchically as shown in Figure 2-5. For example, the term fly-by-wire corresponds to applications in avionics and drive-by-wire covers automotive applications. Within these categories further refinement is possible by naming specific functions like brake-by-wire and steer-by-wire.

All the different by-wire classes can obviously be of the type control system or full system, thus rendering the expression variants *brake-by-wire control system* and *full brake-by-wire system*. In a brake-by-wire control system the information transfer from the brake pedal to the breaking actuator is handled electronically, but hydraulic or pneumatic actuators, probably with electronically controlled valves, perform the actuation. If the actuators are replaced with electrical motors a full brake-by-wire system is created.

2.3.2 Why by-wire?

When a shift in technology is to take place there are some initial efforts that are necessary to make the shift. To overcome these initial obstacles there must be a promise of improvement and return on investment. What makes Drive-By-Wire a winning technology?

There are two main characteristics of drive-by-wire systems that bring along a set of beneficial properties. First, it becomes possible to implement complex control algorithms, adding and improving functions by utilizing the flexibility of software. The electronic control gives benefits in the implementation of more advanced and exact control algorithms at the ECU level. Also, it becomes easier to allow subsystems to work together by information exchange in networks. Second, the by-wire technology facilitates the removal of some mechanical components making the mechanical design simpler and more flexible. Full by-wire provides the full advantage of simpler mechanical design.

Control unit level benefits

Using high-speed electronic braking control, improvements can be made for known difficult situations. The benefits of electronic control for braking have been used in the heavy truck industry for years in order to improve the reaction time of the system. The electric signal is much faster than the pneumatic signal used in older vehicles. The reaction time can also be improved by using brake assist that reacts by preparing braking when the foot leaves the throttle pedal. It is also possible to improve the brake force distribution on the tires enabling more advanced stability control (ESP) functionality [Stoll 2001].

In a passenger car, a comparison was made between a hydraulic system with a brake response of 200 milliseconds per G-force (ms/G), and an electronic system that was able to give a response of 50 ms/G. Using electronic actuators it was possible to reduce the braking distance at 100km/h by 2 meters, from 42 to just below 40 meters. The two systems were also compared in the situation with a rapid change in friction coefficient of the driving surface while braking. The electronic system showed superior performance also in this case. [Yokohama et al 2002].

System level networking benefits

The networking aspect of by-wire systems allows new control functions based on information from separate parts of the vehicle. One implementation that shows how networking can improve performance is the improved braking performance of the '30m car'. A networked combination of sensors in tires, distance sensors and active suspension managed a braking distance of 30.18 m at a speed of 100 kph, approximately 8 m better that a reference production vehicle. Using the technology it was also possible to reduce reaction time and braking build-up and thus improving the entire stopping distance [Rieth & Eberz 2002].

By-wire control or Full by-wire?

Moving over to full by-wire introduces a range of new possibilities in the design of the car. Further, difficult, dangerous and expensive mechanics can be removed together with pneumatic and hydraulic systems. The removal of substances such as oils also provides a more environmental friendly system.

The freedom in the mechanical design provides more available space and the possibility to improve the interface to the driver. This freedom has been well utilized and investigated by several of the available by-wire as presented in the next section.

An important point to make is that a transition to full by-wire technology is necessary for the technology to be cost efficient. Using expensive mechanical/hydraulic backup is not a feasible option for cost-efficient manufacturing [X-by-wire 1998]. The potential economical benefits of a full by-wire system compared to a by-wire control system are enormous. A problem is that it will take a long time to achieve end-user acceptance allowing the removal of backup mechanical and hydraulic components. Due to the acceptance difficulties, a reasonable development is that drive-by-wire systems are first introduced with current solutions as a backup, relying on the benefits of improved control as motivation for their introduction. In the next stage the mechanical and hydraulic systems are slowly phased out one-by-one, allowing users to get aquatinted with the thought of by-wire systems [Scobie et al 2000, Harter et al 2000].

2.3.3 Drive-by-wire case studies and prototypes

By-wire systems are by no means new. There have been several implementations of drive-by-wire systems, both prototypes and commercial systems. Several concept cars with by-wire technology have been designed of which some are covered in [Larses 2003a], a selection of concepts are also discussed below. Further, there are several commercial implementations, mainly by-wire control applications, of which a selection are listed in Table 2-1 and discussed below.

Commercial implementations

By-wire control has been implemented for a long time. Some of the first implementations include braking and throttle control. The first commercial anti brake locking system (ABS) was introduced by Mercedes in 1978. The system was first presented in 1970 but at that time the available microprocessors could not cope with the ABS real-time requirements delaying the commercial introduction by 8 years [Mercedes 2003]. Today, by-wire control is extensively used in many application areas and some full-by-wire systems are becoming available on the market. The commercial implementations referenced in this section are given in Table 2-1.

Table 2-1 Commercial Drive-By-Wire cases

Product/Standard Name	Involved Companies	Year presented	Brake	Throttle	Gearshift	Steer
OBD (On-Board Diagnostics)	-	1988		X		
EDC (Electronic Diesel Control)	Scania	1988		X		
ABS (Mercedes S)	Mercedes	1978	X			
ESP (Mercedes S)	Mercedes	1995	X	X		
Tiptronic	Porsche, ZF, Bosch	1989			X	
Opticruise	Scania	1995			X	
Quadra-steer	Delphi	2000				X
Active Steering	BMW	2003				X

An electronic stability program (ESP) combines by-wire control of braking and throttle to avoid spinning. The technology uses a set of sensors connected to the braking and throttle system. Whenever the danger of a spin is detected, it reacts by selectively applying braking force to the front and rear wheels and reducing or increasing engine torque. [Mercedes 2003]

Brake-by-wire control is used in the electro-hydraulic brakes, labeled Sensotronic brake control (SBC), of the Mercedes SL. The concept has improved the functionality of the ESP and the brake assist functionality, decreasing brake reaction time. Using electronic control it is possible to make the system react to other signals than the actual button or pedal signal. The SBC reacts when the foot is lifted from the throttle pedal and prepares to apply brake force, it also reacts on the speed of foot movement interpreting if the braking is an emergency situation or a regular traffic situation [Stoll 2001].

Throttle-by-wire is the standard way of controlling a combustion process today. Electronic diesel control (EDC), or by-wire throttle, was first introduced in a Scania truck in 1988 and applied to the 470 hp 14 liter V8 engine. Today, the electronic engine control is used on all Scania engines. The control unit is mounted on the engine, which gives predictable conditions and saves space in other sections of the vehicle.

Electronic engine control was first introduced in the late 70s. Today not only control, but also diagnostics, is performed by electronic means. On-board diagnostics (OBD) is even regulated through legislation. The first act was introduced in California in 1988 and is now replaced by OBD 2 legislation from 1993. In Europe EURO-OBD took effect in 2000, a lighter version was introduced in Japan 2000 but requirements similar to Europe and North America is scheduled to be introduced in 2008 [Ogawa & Morozumi 2002].

For transmission the by-wire alternative is automated manual transmission (AMT). With AMT the mechanical connection to the transmission is eliminated and the gears can be chosen by pushing buttons or allowing a computer to run a gear selection program. AMT systems can be added onto existing manual systems, the technology was first developed for motor sports to relieve the driver from using the clutch. (Wagner 2003). AMT was introduced under the name Tiptronic by Porsche, ZF and Bosch on the Porsche 911 in 1989. Today, similar solutions are available from several car and heavy vehicle makers.

Steer-by-wire has been implemented in several prototypes and is seen as a challenge due to the safety-critical property of steering. Steer-by-wire is commercially available in the form of Quadra-steer, four-wheel steering, from Delphi. This concept uses full steer by-wire for the steering of the rear wheels while using conventional steering on the front wheels. With four-wheel steering is possible to improve the turning radius of the vehicle when the front and rear wheel turn in opposite directions simultaneously. This is the case at low speeds, at high speeds the wheels steer in the same direction creating a sideways displacement of the vehicle without turning it, this functionality is intended for high speed highways improving lane shifting capabilities of the vehicle. [Amberkar et al 2001].

A hybrid between conventional steering and steer-by-wire is provided by the BMW Active steering introduced in the BMW's 5-series in 2003. The core of this active steering is a planetary gearbox integrated into the split steering column. The system produces a superposition of the steering wheel angle by an electronically controlled additional angle. Depending on the situation, the feature varies the angle of steering. In situations such as parking the active steering feature makes it easier to maneuver by amplifying the movements of the steering wheel. It increases the steering angle at lower and mid-range speeds, but at higher speeds (e.g. on motorways), the steering angle is reduced. See Figure 2-6 for a schematic overview of the system [BMW 2005].

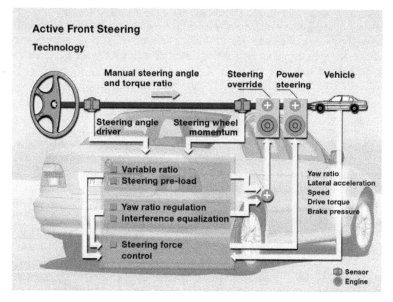

Figure 2-6 BMW Active Steering

By-wire concept cars

Several concept cars are available that utilize drive-by-wire technology to different extents. A selection of four cars, further described below, is provided as examples. The Novanta and Sequel cars are closest to conventional cars while the F400 Carving and PM cars show some possibilities utilizing the freedom of by-wire technology.

Figure 2-7 The Novanta car

The Filo car was designed by Bertone in collaboration with SKF and was presented in March 2001. The Novanta car, shown in figure 12, is the second generation concept car from Bertone/SKF similar to the Filo car. The vehicle utilizes a range of electric actuators developed by SKF and also has an innovative driver interface. The driver has all the vehicle functionality in a special steering wheel, the steering wheel is used for acceleration and braking as well as for steering and gear shifting. The vehicle utilizes a conventional engine for propulsion but electromechanical actuators for braking, clutch and gear shifting.

The Sequel car, shown in Figure 2-8, is the third generation of fuel cell concepts from GM where the idea is to provide a "skateboard" where all the components required for driving the car are included. The first concept was called Autonomy, showed in January 2002 at North American International Auto Show in Detroit. At that time the concept was not drivable but the vision showed a 6" skateboard that was expected to contain all the necessary components for a car. The vehicle was to be driven by a hydrogen fuel-cell and any desirable model could be placed on top of the skateboard. By August 2002 GM revealed the first drivable version of the Autonomy concept, the Hy-Wire car [Hydrogen & Fuel Cell letter 2002]. A comparison of the performance of the second generation Hy-wire and the third generation Sequel cars are shown in Table 2-2, the performance is close to a conventional vehicle [GM 2005].

Table 2-2 Sequel and Hy-wire comparison

	Hy-wire		Sequel	
Thickness	11"	28 cm	11"	28 cm
Power supply	94 kW fuel cell		73 kW fuel cell +65kW Li-ion battery	
Acceleration	0-40mph/64kmh	10 sec	0-60mph/100kmh	9 sec
Top-speed	100 mph	160 km/h	90 mph	145 km/h
Weight	4180 ibs	1900 kg	4780 ibs	2170 kg
Range	60 miles	100 km	300 miles	500 km
Fuel tank pressure	5000 psi	350 bar	10 000 psi	700 bar
Refueling time	5 min			

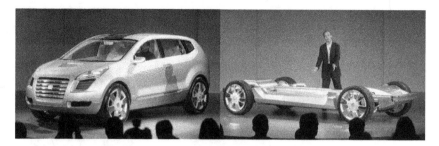

Figure 2-8 The GM Sequel

If the previous cars try to copy the functionality of conventional cars the Mercedes Benz F400 Carving tries to extend the functionality by using the freedom in design allowed by drive-by-wire technology. The F400 have the ability to change the camber angle of the wheels up to 20° which allows better cornering abilities with a maximum side force of 1.28G. The vehicle was first presented at the Tokyo Motor show in late 2001 and is shown in Figure 2-9 [Automotive Intelligence 2001].

Figure 2-9 The Mercedes Benz F400 Carving

An even more extreme utilization of the freedom provided by electronics is shown by Toyota and the Personal Mobility (PM) vehicle, Figure 2-10. It is a one-person vehicle where the driver sits inside a pod-like cabin that is equipped with twin pistol grip controls for by-wire steering, braking and acceleration. Propulsion power is provided from a rear-mounted brushless electric motor. A visual communication system keeps the car in touch with other PMs so the vehicles can share information and "chat" about experiences. According to Toyota, the PMs are social machines and can move in groups (platooning), one taking the lead and the others following at safe intervals on auto pilot. Only the car in front needs to have the destination programmed into its navigation system. The PM also communicates different emotions, using LEDs to display different colors on its body panels, lights and rear wheels, compare Figure 2-10 and Figure 2-11 [Toyota 2005a].

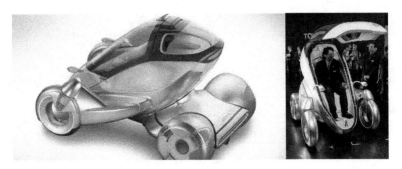

Figure 2-10 The Toyota PM

The angle of the cabin and length of wheelbase adjust according to driving conditions, standing tall for walk-in access, dropping slightly for urban driving and fully extending for high-speed motoring, see Figure 2-11. The front wheels feature independent steering and maneuverability is aided by rear wheels that turn in opposite directions, enabling the car to rotate on the spot [Toyota 2005a].

Figure 2-11 The leaning Toyota PM

2.3.4 Challenges for drive-by-wire applications

So far the benefits and existing implementations of by-wire applications have been discussed. But there are obviously also some challenges for this new technology. The by-wire control requires dependable control algorithms and that the complex dependencies in the network must be managed. Full by-wire must also ensure dependable supply of power for actuators. Further, there are still legal issues that inhibit the use of, for example, steer-by-wire systems. The problems are summarized in Table 2-3 and further described below.

Power/Energy management

One of the main issues that are holding the digital revolution back is the need for more energy. Going over to a full by-wire system does not only remove hydraulics and pneumatics, a lot of energy and power consuming electrical devices are also added. Switching over to a full by-wire system might seem feasible from a control system point of view but the electrical power requirements created by the use of electric actuators are far from trivial. A large amount of energy needs to be supplied with high power. This electrical energy need to be generated and stored on-board the vehicle. Further, the electrical actuators will become bulky and heavy to provide enough force and torque.

Work in this field needs to be done both on the supply and demand side in the power network. Creation and storing of the energy is an important topic but reducing the energy losses in distribution and actuation is not to be forgotten. One necessary change is to increase the voltage in the power net of the vehicle [Steiner & Schmidt 2002]. An increase in voltage decreases current, copper cost, wire dimensions and weight [Frank 2002]. 42V has been discussed for a long time as a new automotive voltage standard after a collaborative industrial/academic effort initiated by Mercedes Benz and MIT. Currently the progress in the development of 42V technology is slow which might be attributed to the economic climate [Kassakian 2002]. Another explanation for the slow development of 42V is the uncertainty in if the standard is right. There are voices advocating that the voltage is too low for future requirements. The 42V standard remains an open issue but vehicles utilizing 42V is coming, both in concept cars, and also in production cars like the Toyota Crown. [Frank 2002]. Simultaneously cars utilizing higher voltages are also developed, for example the commercially available Toyota Prius hybrid vehicle that uses a voltage of 500V [Toyota 2005b].

The challenge of embedded system control complexity

Introducing networked control units running communicating software introduces new complexity into vehicles. The technology is not yet mature compared to, for example, mechanical engineering and there are several points to be proved. According to Feick et al [2000] there are two main fields that should be considered where current mechanical solutions are strong: cost & safety. How by-wire systems can meet these requirements is highly dependent on how the complexity of the network and the software is handled.

Maturity in the engineering process is necessary to test, document and verify the software properly. It is important to have methods to understand the interdependencies within the code so alterations at system level can be performed with little or no risk of malfunctions.

A connected issue is how the hardware running the software should be handled; unless standards are determined vast efforts will go into the specification of hardware. It will also be difficult to change supplier once a working cooperation have been established as there are a large amount of analog and digital inputs and outputs that need to be specified and trimmed avoiding changes at the system level. The combined system of hardware and software must be cost-efficient and dependable across all product variants and for all implemented functions.

Legal issues

Another important problem connected to the reduction of mechanical components is the legal issues. Current legislation inhibits the use of full by-wire systems by specifying technical solutions, two well known examples related to drive-by-wire relate to steering and braking. For steering the use of a mechanical steering column is required by law. For braking, a redundant braking system is necessary, and it is unclear how the redundant braking systems are to be implemented for by-wire systems.

The liability in case of accidents is also unclear, is it the driver or the system that is responsible for the behavior of the vehicle? When new systems are introduced that can cope with bad driving, like electronic stability programs (ESP), it is possible that drivers will rely on the system to take them out of difficult situations. If accidents still occur, is it due to bad driving or due to a malfunctioning system?

Table 2-3 Drive-by-wire benefits and challenges

	Benefits	Challenges
By-wire control	*Advanced control algorithms implemented in software enables improved performance and new functionality such as:*	*Many applications are safety-critical and require verification and validation, complicated by the complexity of software.*
	Active safety - Inhibiting improper driver action - Function supporting the driver *Adaptive/learning behavior* - Improve system quality during use - Adapt to individual driver *Information logging* - Powerful diagnostic tools - Databases for detection of systematic design errors *Improved vehicle dynamics* - Improved braking distance - Networked control algorithms *Efficient utilization* - Engine fuel economy and power - Brake blending - Efficient gear shifting *Freedom in user interface behavior*	*High initial development effort* - Requires extensive initial investment *Increased verification needs* - Increased development time - Increased need for formal methods *Increased system interdependencies* - Co-operation and co-ordination necessary during development - Regulations for updating and maintenance of system - Increased administration overhead
Full By-wire	*Using electrical actuators removes dirty, expensive and dangerous mechanical, hydraulic and pneumatic solutions.*	*Mechanics is a proven technology where engineering is mature and standardization makes solutions cost-efficient.*
	Passive safety - Removal of steering column - Removal of pedals *Reduced weight* - Improved cargo ratio - Improved fuel economy *Environmental friendly "Dry" solutions* - Removal of hydraulics (oil) - Removal of pneumatics (air) *Freedom in mechanical design* - More available space *Freedom in user interface design*	*Legal issues* - Legal requirements on design of brakes, steering etc. - Liability issues *Reduction of redundant mechanical parts* - Decreased fault-tolerance - Reconstruction needed for fail-safe modes

Comparing challenges and benefits

An illustration of the topics discussed in this section is given in Table 2-3 that shows the possibilities and challenges created by by-wire systems. The table separates by-wire control with complex software and full by-wire that enables removal of mechanical components. The challenges are mainly due to immaturity in the engineering of safety-critical electronics while the possibility for functionality is limited only by imagination.

It is possible and useful to distinguish between performance improvements and functional improvements. Performance improvement enhances the performance of the vehicle without effects on functional content, for example through weight reduction. Functional improvement will improve functionality, such as stability control systems or improved flexibility in operator interface design. The benefits of performance improvement are often clear, as the weight reduction. Functional improvements require careful consideration as the market value of a function may be difficult to estimate, it may even be negative if the function is unwanted.

2.4 Networks and architectures

An important challenge for OEMs is to manage the complexity of communication and networks, and to ensure that they run in a reliable and cost-efficient fashion. There are a range of standard protocols commonly utilized for automotive applications covered in this section. Further, the choice of both software and hardware architectures and topologies are important for the properties of the embedded system. Some ideas and standards in this field are also covered.

The broadened opportunities for OEMs within electronics have previously been discussed in relation to the range of automotive applications. Technologies that are in the hands of external suppliers are not addressed here. Typical areas where OEMs should not enter the technical scene but rely on suppliers include TFT-screens, DVD-players and hand-held computers as obvious examples. What is an important issue is to define the border. Interfaces towards consumer products should be defined both for software and hardware and the role of the OEM and suppliers should be clarified. These issues will not be elaborated further in this work.

2.4.1 In-vehicle communication standards

Different buses and protocols for safety-critical applications have been thoroughly evaluated and compared by Rushby [2001]. To be of commercial value, networking must rely on widely spread standards based on standard external interfaces. Standardization of communication is an essential issue that allows further modular development by suppliers [Coelingh et al 2002]. This modular development is necessary for efficiently exploiting development resources in the automotive industry [Malhotra 2002; McElroy

& Goldstein 2002]. This section discusses some current industry standards and future trends of communication protocols.

Time-triggered and event-triggered communication

The protocol for safety critical automotive networks has been a highly debated issue. Much of the discussion has been focused on the approaches to scheduling, but issues on necessary services in the protocol are also debated. The scheduling question is if it is necessary to go from priority-based (event-triggered) systems to time-triggered systems. For safety critical applications time triggered protocols are generally seen as superior to event triggered protocols as a time triggered solution is easier to analyze and more predictable in behavior. Discussing in terms of time-triggered and event-triggered protocols is actually a simplification of the properties of the protocols. Predictable time variation is not only derived from triggering, other factors that influence the timing behavior include scheduling policies (for execution and communication), clock synchronization and the establishment of a global time-base [Törngren 1998]. To simplify the reasoning here a distinction is made between event-triggered and time-triggered protocols.

Scobie et al [2000] conclude that the time-triggered architecture (TTA) can cope with the requirements concerning both cost and reliability. Considering this rather unanimous theoretical vote for time-triggered protocols it is interesting to see that practical implementations are prone to use the event-triggered CAN protocol. An example is the implementation of sensotronic brake control (SBC) in Mercedes SL-class cars. This is an electro-hydraulic brake-by wire system, utilizing CAN for communication [Stoll 2001]. The obvious reason for this is that the event-triggered CAN network is more or less the only standard that has been available for a long period of time and can qualify as a proven technology.

The need for a time triggered solution is not because of requirements from a single function, instead it is required to enable dependable system level integration of functions. Time-triggered protocols are deterministic and communication patterns can be verified. Each application needs to be separated from the others ensuring non-interference. Currently the complexity of systems can be handled by making sure that enough bandwidth is available, but increased use of software will change the requirements on the in-car network. It is probable that different applications will use different protocols, and safety-critical applications will use a deterministic time-triggered protocol [Kopetz & Grünsteidl 1993].

There are a range of services and mechanisms that different bus protocols employ. Essential basic services include clock synchronization, time triggered activation and reliable message delivery, but there are several other services that can improve the dependability of the communication implementation covered by Rushby [2001].

A multitude of protocols

Many different communication protocols have been proposed for vehicular purposes. It is a general notion that due to varying requirements on the communication for different systems, several alternative protocols will be used simultaneously. For cost minimization the vehicle will contain separate networks connected through gateways, the cheapest solution meeting the requirements for a specific network will be used. Using different subnets is a natural path that also provides partitioning and separation of concerns. Using several different protocols increases the cost for tools and maintenance. To reduce costs and to easily achieve gateway functionality, it is desired that all the protocols are standardized [Emaus 2000].

Table 2-4 Comparison of common communication protocol characteristics

	Transmission media	Bit Rate	Cost	Max bus length	Max nodes
CAN J-1939	Twisted pair	250 kb/s	Medium	40 m	30/10
CAN J-1850	Twisted pair	125 kb/s	Low	35 m	32
CAN 2.0B	Twisted pair	1 Mb/s	Medium	40 m	32
LIN	Single wire	20 kb/s	Low	40 m	16
Flexray	Optical/Wire	10Mb/s	Medium	?	?
TTP	Optical/Wire	5-25Mb/s	High	?	64/256
Firewire	Shielded twisted pair	98-393 Mb/s	Medium	72 m	27
IDB-1394	Optical	98-393 Mb/s	Medium	Infinite	63
MOST	Optical	25Mb/s	High	Infinite	64

The Society of Automotive Engineers (SAE) has categorized bus protocols into class A, B and C. Where class A is used for low-end general purpose communication and class C is used for safety-related real-time systems. Another four classes have been suggested to include buses in the categories of diagnostics, safety, mobile media (low speed, high speed and wireless) and X-by wire. The trade off, placing protocols in different classes, is between low cost and dependability and there are a variety of protocols available in any category [Lupini 2001; Lupini 2003; Quigley et al 2001]. Also, the requirements differ between the need for a deterministic protocol for safety-related control mechanisms and the need for real-time streaming of media from entertainment applications or cameras around the vehicle supporting the driver. There is a set of existing and currently developed automotive standards implicitly related to these classes, the characteristics of a selection of commonly referred standards are given in Table 2-4 [Lupini 2003; Teepe et al 2002].

CAN and J-1939 (class B and C)

The most established of all standards is probably the controller area network (CAN) that has been used in cars since the early 90s when it was first introduced by Mercedes. CAN is also utilized by a set of higher-level protocols, like J-1939 and Volcano for automotive control, and J-1850 for diagnostics. J-1939 is the truck and bus standard protocol. The CAN-protocol can also run at different speeds to a maximum of 1 Mb/s, 250 kb/s is specified for J1939. Due to the maturity of the protocol CAN-based components are easy to find at a fairly low cost.

LIN (class A)

The LIN-bus is a low-cost, low speed serial communication bus for low-end applications. It belongs to the SAE class A buses and is currently the standard bus in this class. It is usually used for simple switches, door electronics, seat controls and similar applications. The LIN bus uses a Master/Slave approach, having one Master and one or more Slaves. The LIN bus does not need to resolve bus collisions because only one message is allowed on the bus at a time.

TTP (class C, by-wire)

The Time-Triggered Protocol (TTP) was designed for safety-critical high-speed applications. In contrast to classical event-triggered communication systems, the Time-Triggered Protocol involves a continuous communication of all connected nodes, for example steering wheel or brakes, via redundant data buses at predefined intervals of microseconds. The communication is statically scheduled which ensures that an overload in the bus system is prevented even if several important events occur simultaneously, e.g. over-steering the vehicle and braking at the same time. All events are safely processed according to schedule without data collision. The TDMA (time division multiple access) bus access scheme is collision-free.

The communication controllers available today support 25 Mbit/s synchronous and 5 Mbit/s asynchronous transmission rates (asynchronous transmission is the method used over twisted pair wiring, synchronous transmission uses Ethernet-like wiring). Prototype implementations have used 1 Gbit/s technology (lab experiment in 2002). TTP networks can contain up to 64 nodes. The cabling topology can be bus, star, or any combination of the two. Multiple stars or sub-buses on stars are also supported.

TTP can only send time-triggered messages, but applications using TTP can use event-triggered messages. The transmission of event-triggered messages is performed over an event channel (bandwidth is reserved for event transmissions inside the TDMA slots, and the messages use identifiers). A typical event channel mechanism is a CAN emulation, in which a CAN-compatible interface is provided and the CAN messages are transmitted inside TTP data frames. Allocation of bandwidth is statically predefined per node and arbitration is not performed among different nodes (as in CAN or Byteflight) but only among different functions within a node. Timing and bandwidth analysis for event transmissions is therefore done on a per-node basis and does not need system-level design [TTTech 2005].

Flexray (class C, by-wire)

FlexRay is a high-speed serial communication protocol for in-vehicle networks that combines time triggered and event triggered messaging. The protocol supports bus, star and multiple star topologies. The FlexRay bus specification provides an electrical and an optical version of the physical layer. FlexRay is a fault tolerant bus and provides deterministic data transmission at a Baud-Rate of between 500kb/s to 10Mb/s with a 24 bit CRC. The 10Mbit/sec data rate is available on two channels, giving a gross data rate of up to 20Mbit/sec. FlexRay is an extended protocol version of byteflight, developed by BMW [Flexray 2005].

MOST (high speed media)

MOST is a standardization project in the high-speed mobile media class for multimedia applications with a transfer rate at 25Mbit/s. MOST is currently competing with the Firewire (IEEE 1394) standard in this class. MOST is implemented as a peer-to-peer point-to-point network which can be implemented in a ring, star or daisy-chain topology. The protocol is based on optical fibre communication.

The design approach behind MOST technology is to provide a low overhead, low cost network interface to even the simplest multimedia device. It supports devices with low intelligence and no buffering capacities such as D/A converters for speakers as well as much more complex, DSP-based devices and their need for sophisticated control mechanisms and multimedia capabilities. [Mostcooperation 2005]

Firewire and IDB-1394 (high speed media)

The current Firewire (IEEE 1394a standard) bus transmits data packets at rates of 98, 196, and 393 Megabits/s, designated S100, S200 and S400. While the standard permits up to 27 connectors, a typical Firewire device would have three connectors, to allow it to be used in a tree topology with one parent and two child nodes. However many current devices have only one or two connectors, for use as end nodes in a chain, or as individual nodes in a daisy chain.

The Gigabit Firewire or IEEE 1394b standard is being developed, with operation over distances of 50-100 meters and at burst data rates of 800 - 1600 megabits/s, 3200 megabits/s being planned in as a growth feature. The new standard supports operation over shielded and unshielded twisted pair as well as optical fibre.

IDB has developed a version of Firewire for automotive applications, the IDB-1394. This protocol is built on IEEE-1394 with an optical media and is designed for high speed multimedia applications that require large amounts of information to be moved quickly on a vehicle. The IDB-1394 will support speeds of up to 400Mbps (S400), and support devices operating at S100, S200, and S400. The number of devices on the bus is limited to 63 nodes. [IDBForum 2005]

Under development is IDB-1394Cu, a copper-based solution. Other future versions of IDB-1394 on alternate physical media are being considered for automotive and specialty applications. The optical version of IDB-1394 is expected to be introduced in models in Japan and Europe throughout 2006-2007 [Freeman 2004].

Optical media

The traditional glass fibre optical media has faced problems for automotive applications by being to costly and difficult to connect. A potential solution is POF (Polymer Optical Fibre) that provides cheaper connections than optical glass fibre. The cost for each POF connection is approximately 1/100th the cost of a glass fibre connection. This is because polymers are more flexible than glass allowing POF to be manufactured with larger diameters. Polymethyl methacrylate (PMMA) is a fibre type that offers low attenuation and can handle 10 Gbit/s data rates. But PMMA fibre still has one key limitation, it can only be used at temperatures below 85 °C. For cars significantly higher temperatures can be reached. Temperatures in a roof module can go above 100 °C, and 125 °C is not unusual in the engine compartment. As a result, it is desired that the fibre is specified to operate at temperatures of 125 °C [Freeman 2004].

One solution to the temperature problem is to use a hybrid fibre called polymer-clad silica (PCS). PCS fibre offers many of the advantages of POF including large core size and the associated ease of connection, but it is also capable of high-performance operation at up to 125 °C. The combination of silica core and hard polymer cladding offers the best of both worlds, incorporating low-loss transmission and superior fibre strength [Freeman 2004], if the media will be cost-efficient in automotive applications remains to be proven.

Future dominating communication/industry standards

Systems' architectures are predominantly determined by the chosen technology platform and its inherent properties. This makes it very important to monitor and define future industry standards that will be the basis for future system architectures. For the communication standards, there is currently a general opinion on what the most probable standards that will join CAN will be in the automotive industry. Two very probable new standards are LIN and Flexray. The LIN bus is already seen as the standard solution for simple low-cost applications. For by-wire applications it is seen as required to move towards a time-triggered solution. The main contenders for the by-wire segment have been Flexray and TTP and some discussions have existed on the option to add features to CAN and create a time triggered CAN (TTCAN) [Koopman 2002]. Flexray seem to win the race as most of the major car manufacturers are currently supporting the protocol, for example does BMW intend to replace all CAN-based communication in chassis systems with Flexray.

The MOST consortia started out with strong support in the industry and may be the protocol of choice for infotainment applications but this is yet to be proven. Carmakers have problems with the protocol based on the optical media, running into problems as they implement MOST networks in series production. General Motors have stated that they will not apply MOST beyond the Saab 9-3 [Hansen 2004]. Firewire (IEEE-1394) and the related IDB-1394 may be a contender considering the increasing use of Firewire in consumer electronics.

Within the SAE safety class (not the SAE C-class), intended for airbag and similar applications, there are still no natural standard and many buses are proprietary [Teepe et al 2002]. The standards for wireless mobile media are treated in the section of telematics.

2.4.2 Telematic solutions

Telematics originally referred to the broad industry concerning the use of computers and telecommunication systems together. Today, the term has evolved to refer to wireless automotive applications including GPS and general telecommunication functions that originate or end inside the vehicle [Webopedia 2003]. Telematics is a combination of communication technologies, in-car information systems and in-car computing. Ranging from hands-free functionality to dynamic navigation and location based services [Ender 2002].

There are several possibilities to achieve wireless communication in vehicles. One distinction is between wide area networks (WAN), generally implemented by cellular networks, and wireless local area networks (WLAN), commonly implemented by the standard IEEE 802.11b. There are also personal area networks (PAN) for short-range communication that can be considered to be similar to WLANs for some applications. Bluetooth belongs to the PAN category [D'Avello & Van Bosch 2002].

Another category that only achieves one-way data transfer is broadcast distribution systems (BDS) similar to the traditional car radio. The BDS systems are using digital transmission and both terrestrial and satellite solutions. The different types of wireless solutions and common implementations are given in Table 2-5 [DeVries et al 2002].

Table 2-5 Telematic solutions

Acronym	Name	Implementations
WAN	*Wide Area Network*	GSM (Global System for Mobile-communications) CDMA (Code Division Multiple Access) GPRS (General Packet Radio Service) UMTS (3G) (Universal Mobile Telecommunications System)
WLAN	*Wireless Local Area Network*	IEEE 802.11x HiperLan
PAN	*Personal Area Network*	Bluetooth IEEE 802.15.4
BDS	*Broadcast Distribution Systems*	SDARS (Satellite Digital Audio Radio Service) DAB (Eureka/147 Digital Audio Broadcasting) IBOC (In-Band On-Channel

Some applications are geographically global and are utilized wherever the vehicle is and thus require WAN or possibly BDS solutions. Other applications are local, only applicable for a specific geographical region or place, and these may benefit from solutions with WLAN or PAN. Today connectivity is usually solved through cellular communication but in response to the different needs, mobile gateways with a combination of cell phones and WLAN technologies are seen as a possible solution [D'Avello & Van Bosch 2002; Kanayama et al 2002]. The different available technical solutions are considered to be working well in terms of providing communication as stand alone applictions, the remaining question is how the solutions are to be properly integrate into the vehicle.

As telematics of today are based on cellular phones the approach for integration of this component becomes a central issue. The question is if the phone, or cellular device, should be embedded in the vehicle, portable or a combination of the two [D'Avello & Van Bosch 2002]. Embedded devices are closely linked to the vehicle while portable devices are closely linked to an individual. Back-seat applications and personal services that only use the car as conduit suggest that the cellular devices should be portable. Vehicle-centric applications on the other hand, that use the car as an end, would suggest embedded connectivity. Typical vehicle centric applications include safety and security like emergency services and stolen vehicle tracking. Another argument for non-embedded devices is the problem with upgradeability. Phones, like other consumer electronics, have development cycles of months before a new product arrives. Vehicles have cycles of years. Embedding a specific device might mean that the cellular device is hopelessly outdated when the vehicle hits the market. Using connectivity through technologies like Bluetooth allows the consumer to choose the latest phone technology and integrate it with the car [D'Avello & Van Bosch 2002]. Using both approaches is a third possibility that is more expensive but enables more services and more reliable services. It all comes down to the integration problem of choosing the right composition of applications. There are several available solutions and they are all working.

"The technologies are ready for the applications, but the applications may not be ready for the technologies." [DeVries et al 2002].

2.4.3 System architecture trends

The system architecture of automotive electronics has evolved over time. In the 90s the EE-architecture developed organically bottom up, adding a new control unit for every new function introduced. This development method is not sustainable in a future scenario with constantly increasing number of functions. Both the complexity and costs must be kept at bay, solutions require both an approach to manage the network topology and also a strategy to reduce the number of control units to reduce hardware costs. Hans-Georg Frischkorn, Senior Engineering Vice President for Electric/Electronics at BMW pronounced a strategy in a keynote at SAE World Congress 2003 to halve the number of control units in five to ten years. Meanwhile architectures are needed that can cope with the high number of control units.

Topologies, ECUs and three levels of concerns

In response to the growing networks means to handle the complexity in them have been developed. It is well known that when complexity increases the necessity for new levels of abstraction increases [Shaw 1989]. In the automotive industry both the physical network topology and the role of ECUs are changing. Multiple networks with different protocols responding to different requirements are emerging [Emaus 2000]. These networks are becoming arranged in hierarchies with sub-networks defined at vehicle system level [Steiner & Schmidt 2001].

Because of the many systems that need to be integrated is it not feasible to use a single centralized 'super gateway' in the long perspective. The complexity of the software in a central gateway will increase with the amount of functionality supported, the number of communication protocols that needs support, as well as configuration issues where different products require different gateway specifications depending on the configuration. This increased complexity will reduce the reliability of the component and jeopardize the complete system as the central gateway is a single point of failure.

As an alternative a backbone-architecture with decentralized gateways should be used, this is a more modular and flexible design that can handle product variety. The backbone is a bus dedicated to integrate the other buses, using this architecture gives a scalable solution where there is less need to change hardware if the network is to be expanded, instead another subnet can be attached to the backbone. The integration challenge is to find a good backbone-architecture and define the interfaces to the separate systems [Beck et al 2001]. Today a division into three networking levels is suggested: Backbone & Gateways, Cluster networks and Local ECUs with mechatronic sub-modules as illustrated in Figure 2-12 [Reilhac & Bavoux 2002].

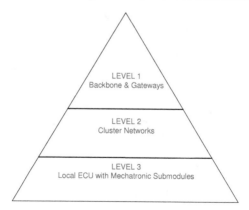

Figure 2-12 System design levels [Reilhac & Bavoux 2002]

The three levels each represent an epoch of the development of automotive electronics, adding new levels in the hierarchy. First analog systems and wiring were connected to separate and independent ECUs, and then the ECUs became connected in multiplexed networks. Today the networks have grown and must be separated by gateways and integrating backbones.

Using this high level view it is possible to break down an architecture and address different parts of it separately. At the top level the different sub-networks and their responsibilities are defined. An example of suggested sub-networks include powertrain, chassis, body/comfort and telematics where each can utilize a different protocol suitable for the local needs [Beck et al 2001]. Using separate but integrated networks allows good control over the system, it is possible to allow systems of high criticality to be interconnected to systems of low criticality by using strictly defined interfaces or firewalls [Watt 2000]. How the communication between objects with different criticality is to be handled is an area with ongoing research. Classic policies inhibit information flow from low-level to high-level criticality objects, newer policies work with authorization in the flow between objects of different criticality [Totel et al 1998].

At the second level, the internal topology and the level of distribution of control within the cluster networks need to be decided, this is closely related to how the role of the ECUs is defined at the third level. A problem here is the issue of mixing components with different criticality on shared resources in a system, like software components sharing a processor. This requires that non-interference between the objects of different criticality can be assured [Dutertre & Stavridou 1999].

Distribution of control in cluster networks

Within each defined sub-network a topology for the distribution of control is needed. It is possible to classify architectures designed for a specific function or a set of functions. Five basic classes of structural architectures can be distinguished as illustrated in figure 18: Independent, centralized, localized, modular and smart architectures. Traditionally, architecting has been aiming for one of these to be predominant in the entire network or several cluster networks. Newer ideas suggest that each function need to choose its optimal architecture based on the specific requirements of that function [Reilhac & Bavoux 2002]. When several functions will share hardware the different topology requirements become an integration issue.

Independent functions are implemented through dedicated cabling and control and without sharing any resources with other functions.

Centralized functions are implemented in a single ECU, integrating control and enabling interactions between the functions through software implementation. Dedicated cabling for power distribution and analog signals are used.

Localized functions use local control units for control and power distribution through these control units at the local position. Information is shared among the local control units over a data network. A localized architecture reduces the wiring compared to the centralized architecture.

Modular architectures are built by placing regional control units that communicate over the network. These regional units are chosen in order to ease diversity management. Compared to the localized architecture, a modular architecture is tailored to fit into the modular program of mechanical components.

Smart architectures use plug and play components down to sensor and actuator level. Each sensor and actuator component are directly connected to buses that are connected to generic control units for plug and play functionality

Figure 2-13 Architecture options

Traditional architectures have been independent lacking both means and reasons to be integrated with components implementing other functions. With increasing electronics content independent architectures become costly and lack the benefits of networking. Today systems are moving over to centralized and localized architectures where functions are implemented in single dedicated or a few ECUs to reduce hardware costs. The current trend is towards even more distributed and connected architectures where the functions become more integrated. [Teepe et al 2002].

The trends in topology and control distribution in automotive networks show the underlying functional structure. Earlier functions were independent, and thus the control system was also built from independent components. Today, dependencies are becoming more and more complex as different applications like adaptive cruise control and stability programs requires the services of several subsystems. Some of this complexity is handled by building hierarchies, hiding sub-systems.

Some kind of connected architecture is inevitable for by-wire systems, but other unrelated systems in the vehicle may still be independent. For by-wire systems centralization is not a possible option, the real-time requirements and the reliability requirements are hard to meet cost-efficiently if a single powerful processor is used. Current ideas imply a layered system design where higher order functions are implemented in more central nodes and lower level control is performed locally. An

Architecture trends
Past
Functions: Independent
Topology: Separate ECUs
Today
Functions: Dependent
Topology: Networks
Future
Functions: Layered
Topology: Hierarchies

example is given by the complete vehicle control (CVC) architecture that uses a centralization of core functionality in order to have a more clear system interface from driver to vehicle [Coelingh et al 2002].

If the system is arranged in a layered structure with defined interfaces the complexity can be managed simultaneously as flexibility will be available. It will be easy to add and remove functionality as well as exchanging physical components with limited alteration within the control system. With a more distributed control solution separating between high level applications and low level local control loops a probable development is towards client/server architectures. The clients in this architecture would be based on the old type of ECUs and implement real-time and safety features, including local control algorithms, ensuring that a certain degree of service is always available. The servers, based on a more generic hardware, implement the higher order services with improved functionality [Topp & Weber 2001].

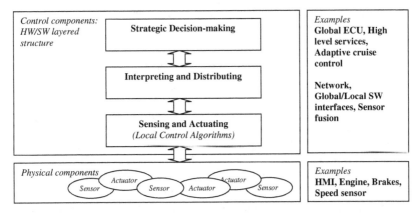

Figure 2-14 Layering of functionality (based on Steiner & Schmidt 2001)

Using more distributed architectures a separation of concerns into layers may be beneficial, increasing the possibilities for reuse of qualified parts of the system [Teepe et al 2002]. In a layered structure communication is only allowed between adjacent layers, which means that the context of a specific software block becomes bounded and possible to document. A separation between low level control and higher order control layers can be used, as illustrated in Figure 2-14. The low level control is then decentralized while higher order control is centralized on standardized control units [Steiner & Schmidt 2001]. Within these major layers it is possible to define sub-layers according to a functional decomposition, structuring the software further.

2.4.4 Standardization beyond communication protocols

The more relationships being introduced across units, the more the system becomes reliant on systematic design to manage the complexity in these relationships in a cost-efficient manner. One contribution to the solution of this complexity management issue is the definition of a set of open standards. If the different actors, such as OEMs and suppliers of SW, HW and application data content, shall be able to profit from cooperation it is necessary that the boundaries of interaction are established. This requires standards. Standards have to be declared and accepted in many fields: communication protocols, operating systems, voltage levels and development tools just to name a few areas.

With open standards, compared to proprietary architectures, development efforts can be divided among several actors. One possible future scenario is that hardware control units together with diagnostic software and some other software components become standardized commodities. However, with a standard the exclusivity of systems is also lost. All the competitors will have access to the same technology and efforts to improve a sub-system will contribute to all the players in the industry [Würtenberger 2002]. Engineers must excel by building the applications on top of the standard components. This may not be a problem as car and truck manufacturers do not traditionally distinguish themselves through diagnostics but rather through the tuning of the vehicle in terms of system design and critical components like control algorithms. The standard also allows new competition to enter the market increasing the price pressure on suppliers. Higher volumes of standard products would probably also contribute to lower costs [Coelingh et al 2002].

A possible alternative in some areas is to solve the integration through process or organizational means, redefining the role of development teams to fit the complex development process of today, both within the OEM organization and also in the organizations of suppliers. These options are further elaborated in chapter 4.

However, using standards brings in an overhead cost for resources in the implementation, as non-standard systems can be tailored to have a more efficient infrastructure. This has traditionally been an inhibiting factor for real-time systems as the requirements on these systems have pushed for extreme optimization of performance under tight resource constraints. Improved hardware and software for lower costs are relaxing the need to optimize the implementation but the real-time requirements will heavily influence any adopted standard.

An earlier standard that has been very successful is the CAN communication standard, and also the J-1939 standard for truck and bus CAN networks. Even if these standards are far from perfect and the restrictions and limitations in the standards, specifically the J-1939, are slowly becoming unbearable for contemporary architectures, they have supplied a common platform enabling rapid development of compatible functions and components. Around the CAN-standard a community of tool developers, hardware suppliers and applications has evolved beneficiary for all the involved parties. With a standard the interface to the "outside world" is fixed and development and exchange of components is simple. It also enhances system integration possibilities for flexible system configuration and reuse of tested building blocks [Würtenberger 2002].

For telematic applications the use of open standards is an obvious solution. The internet has already evolved through open standards and connecting the vehicle to the internet standards brings the entire power of the web into the vehicle. Searching for proprietary solutions here seems like a vain idea, backseat applications require plug and play interfaces. How a vehicle fits into the web is another issue to ponder, but the use of open standards in this area is inevitable [Loose et al 2002].

Current efforts

One group working with standardization for streaming media and external connectivity is the Automotive Multimedia Interface Collaboration (AMI-C). This organization does not develop standards itself but rather adopt existing standards on the market suiting automotive needs, and also communicates automotive requirements to developing standards. AMI-C is currently engaged in the work with Bluetooth, MOST and IEEE 1394 for example. The organization is also working on a vehicle gateway architecture with access management, thus working on the issue of external connectivity [Malhotra 2002]. Recently Nissan released a prototype vehicle compliant to AMI-C specifications based on IEEE 1394 standards [AMI-C 2004].

Another effort aiming at the automotive embedded system is the EAST-EEA (Embedded Electronic Architecture) research project [EAST-EEA 2005]. This project involved European automotive manufacturers and related suppliers and tool vendors as well as academic partners. In the project a layered software architecture based on a middleware concept was implemented. The middleware offers API services to the application layer that allows transparent mapping and interactions between different application functions in the vehicle. The communication layer offers basic communication services to the middleware that can be adapted to the vehicle networks via device drivers.

Related to the EAST-EEA architecture is a process as well as an architecture description language (EAST-EEA ADL) defined to support the development process. The development and validation of architecture and functional modules is another building block within EAST-EEA. The complete development process from specification, simulation, implementation to functional and integration testing is taken into account.

An industrial effort influenced by the results of the EAST project is the development partnership AUTomotive Open System ARchitecture (AUTOSAR), formed by a broad group of players in the automotive industry in 2003 [Heinecke et al 2004]. The technical concept of AUTOSAR is to provide a common software infrastructure for automotive systems of all vehicle domains based on standardized interfaces. The architecture is illustrated in Figure 2-15.

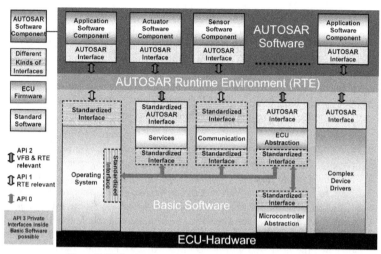

Figure 2-15 The Autosar software architecture [Autosar 2005]

Software components are allocated to a specific ECU and also include special sensor and actuator components, also allocated to ECUs. In the architecture, software components are interacting through a Run-time Environment that implements a Virtual Functional Bus (VFB), as illustrated in Figure 2-16. The run-time environment is running on top of a Basic Software including diagnostic software, operating system and communication drivers. The technical architecture is also accompanied by a design methodology and process provided by AUTOSAR [Autosar 2005].

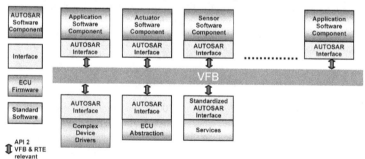

Figure 2-16 The Autosar Virtual Functional Bus (VFB) [Autosar 2005]

2.4.5 Process measures to cope with increasing complexity

To cope with the complexity of a networked system architecture, cost-efficient and structured development methods need to be defined [Reilhac & Bayoux 2002]. The evolution to find these methods has been progressing at most of the companies in the industry.

BMW recognizes that systems architecture design is mainly driven by the collaboration of experienced developers in the different vehicle domains with some central architects moderating discussions. The process is bottom-up oriented, participative and lacking formalization. This approach favors locally optimized solutions, proprietary concepts and risk aversive strategies. To achieve systematic and strategic decisions a normative and model-oriented procedure with a top-down perspective must be introduced to complement the existing procedures [Reichart & Haneberg 2004].

Toyota is working with quality improvements supported by the Capability Maturity Model. Standardization and reuse of general-purpose software components are seen as key issues together with improved specifications that are attributed more than 50% of the system quality [Shigematsu 2002]. DaimlerChrysler describes their process as having gone from a reactive phase with little concern for EE issues, to a design/test phase where the problems were acknowledged and designed for at component level. Today they claim to be in a systems phase designing the architecture and interactions among components, the next step will be to further understand the system level and the real world in a stochastic phase [George & Wang 2002].

Besides design processes incorporating system level concerns there are efforts to find improved topologies and hardware structures that can support the complexity management. A common conclusion is that components and interfaces need clear definitions and standards in order to make the system level design easier, standardization efforts related to the development process are ongoing as previously described [EAST-EEA 2005, Autosar 2005]. Clear boundaries and interfaces are important for future development in dispersed organizations. More intimate relationships with fewer but larger suppliers are assumed to be necessary to maintain product quality, distributing some integration responsibility down to the suppliers. Such an organization can only be enabled by a structuring of the control system network around a set of easily maintained and standardized interfaces.

2.5 Architecture in the SAINT case study

In the SAINT case study [Blixt et al 2005] the goal was to develop a configurable demonstrator in the shape of a scale model truck. Pictures of the final demonstrator are shown in Figure 2-17. The demonstrator was configured through an external system and to allow easy reconfiguration and change management of the software, a product architecture with a middleware similar to the Autosar and EAST-EEA proposals was developed and implemented.

Figure 2-17 The SAINT demonstrator - chassis of the SAINT truck and trailer

2.5.1 Hardware platform

The complete system has six ECUs, three mounted in the truck and three in the high-tech trailer. The six nodes are interconnected with a CAN-bus running at 1 Mbit/s. A WLAN/CAN-bridge is also connected to the bus communicating with CAN wirelessly to an external PC that constitutes an operator station. See Figure 2-18 for a conceptual overview.

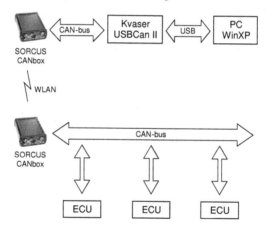

Figure 2-18 Basic structure of the hardware platform in the SAINT project

Several sensors are used to deliver data to the ECUs. For example, a distance sensor based on laser and two ultrasonic sensors are detecting objects in front of the truck. The motor is equipped with an encoder to measure the axis angle. The trailer includes weight sensors measuring both the total load and the distribution on the trailer axles. Other sensors include a vibration sensor and micro switches connected to the alarm. The truck is running on 24V in order for the electrical motor to deliver enough power with lower currents. The hi-tech trailer is running on 12V.

2.5.2 Software platform

In the project the ambition was to create a software and hardware platform that enables the creation of modular applications that can be reused regardless of changes in the hardware architecture. The software applications become hardware independent and the actual functionality of the software is in focus. To ensure this flexibility some system software that is a part of the platform is required, similar to the *basic software* of Autosar mentioned in the previous section. Without such software the implementation will be strictly bound to the hardware target. The middleware implemented in the SAINT project is based on the RTOS OSE Epsilon from Enea. Several services provided by the RTOS were utilized, for example:

- *Communication between processes on the same processor*
- *Scheduling of applications*
- *Priorities of applications*
- *Fault management*

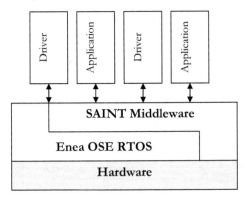

Figure 2-19 The SAINT platform

The middleware manages the communication between the application software components. The RTOS is used by the middleware to provide a range of basic services; the middleware itself extends and adapts the provided basic set of services. The middleware registers all data that is available on the ECU and manages the communication between different software tasks internally as well as distributing required data to other ECUs via the CAN-bus. The software components can be arbitrarily allocated and the middleware will make sure that they are able to communicate.

In the project the functions were modeled as UML 2.0 activity diagrams. The activities in the model were mapped to the application software and were divided in drivers and applications. The drivers are bound to the I/O ports by reading sensors and activating actuators. The drivers convert physical signals from sensors to software variables, and the other way around for actuators.

The applications contain functionality such as control algorithms and logic for activation and deactivation of functions.

The middleware also manages system parameters. These parameters provide configuration support for configurable functions by selecting the processes to be executed in the run-time system. The parameters also provide numeric values such as the value for the speed limiter in the truck. The parameters are stored in a separate memory (EEPROM) that is easily accessible from the operating station via the CAN-bus.

Further, the middleware detects delays in the communication and generates timing errors when a time-out occurs. The middleware contains procedures to detect if the missing value is critical or not, depending on the type of variable that caused the time-out.

2.5.3 Models to support the lifecycle process

All user functions in the system were modeled in activity diagrams where the activities in turn are implemented in software either as application software or drivers. Applications are pure software components and only communicate with other software. The drivers also communicate with external ports to read sensors and activate actuators. Several user functions can utilize the same activity. This model structure modularizes the source code and allows easy configuration of the system. The activities communicate through dataflows modeled in the activity diagram. These dataflows are managed by the middleware in the final implementation.

The middleware utilized in the SAINT project provides a user friendly API for the software developer. It is easy to provide configurability through parameters and it is also easy to build modular software that can be configured before compilation. The applications are integrated through the middleware and the RTOS which make the application source code hardware independent. Adaptions to new hardware are performed in the middleware and RTOS layer. The software is easily moved across control units in order to even out the resource utilization.

With the modular design of the software, change management is much easier. The interface for a given software component is well defined and the results of removing or changing a component can easily be analyzed. The software architecture provides support for configuration and re-configuration at the design, production and service stages in the product lifecycle.

2.6 Concluding remarks

The core of future innovations will be based on software and electronics. As presented in this chapter, electronics is used and will be used for a variety of applications in the automotive sector. Some of the applications are safety-critical and concern the driving of the vehicle which enhances the need to provide dependable solutions. To achieve increased functionality, control units are connected in networks and sensors and actuators are shared across the system by a range of distributed functions implemented in software. To cope with the communication needs in the network, and also to reduce complexity by performing some integration in local networks, one trend in architecture design is to utilize hierarchical networks. The local communication networks are connected by more powerful communication backbones with a few central control units.

Two architecture strategies for decentralization and treatment of software in the architecture can be identified. Either a functional decomposition strategy is applied aiming at a hardware modularized architecture where a separate hardware carries each function, including any software required by the function. Or, a centralization strategy (of software) is adopted, where software is integrated into fewer and more powerful control units and functions are distributed in the network. The traditional approach is to rely on functional decomposition.

With a chosen product architecture it is necessary to define a related development process supported by a proper organization that can be used for efficient continuous development, the architecture must be scalable. The traditional approach, where a new control unit is added for each new function, provides an easy way of distributing work in an organization and arranging a simple and straightforward design process based on control unit design and testing, followed by system integration and testing. With a modular hardware architecture, structure related costs can be reduced. However, the resulting architecture is hardware intensive and not volume cost-efficient, this will be further discussed in the next chapter.

If the strategy follows the current trend to reduce hardware costs by integrating functions and software in fewer hardware units, the organization patterns and development processes become more complex [Reichart & Haneberg 2004]. Then it is not possible to define a general platform in hardware alone. Architectures must also include a communication and software architecture and a related process that allows functions to easily be added, modified and removed. Communication standards and protocols exist, and efforts are in progress to provide more standardized software frameworks. The idea is to define industry wide open standards for the basic infrastructure of the system architecture while competing on the implementation of functions.

3 Background - Basics of Cost and Dependability

This chapter develops some views on the requirements on automotive electronics. Cost-efficiency and dependability are discussed. Cost can be derived from several sources and allocated to products in a variety of ways. Dependability is defined as a composition of several properties, of these are reliability, safety and maintainability given specific attention. This section discusses aspects of cost and dependability in general and also in an automotive context.

The chapter is based on [Larses 2003b].

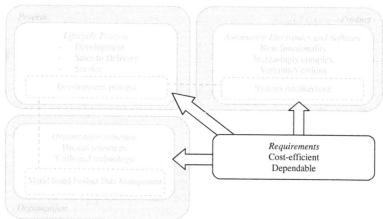

Figure 3-1 Chapter focus - Requirements

3.1 Cost and cost-efficiency

Innovations in the automotive industry today are strongly related to electronics. More than 70% of innovations are software based [Steucka 2003], numbers as high as 80-90 % are mentioned by some sources [Steiner & Schmidt 2001; Knippel & Schulz 2004]. The production cost portion of electronics in a vehicle has been constantly increasing, from less than 5% in 1975 to 20-30% in a regular vehicle today, even more in luxury cars [Teepe et al 2002]. The cost of electronics in the total cost of the vehicle is 22% today and it is expected to be close to 40% in 2010 [Altera 2005]. This implies that cost management of electronics have an increasing leverage on the total cost of the vehicle and these cost must be managed not only in the product, but also in the process and organization for the complete vehicle lifecycle.

Cost and cost analysis are well known in business administration literature. To provide a general background some basic ideas, models and definitions from an introductory book in business administration [Ljung et al 1994] are summarized here. First we must define cost.

Cost. *Cost is consumption of resources.*

Resources are provided by an organization and it is not always straightforward to measure the consumption of resources related to a given product. If costs can be traced to a specific product they are labeled separate costs, if not they are labeled common costs.

Separate cost. *Separate costs are costs that are traced to a specific product or service.*

Common cost. *Common costs can **not** be traced to a specific product or service.*

Because of the problematic nature of tracing costs it is useful to make a further definition. For a specific product it is possible to distinguish between direct and indirect costs.

Direct cost. *Separate costs that are easily measured.*

Indirect cost. *All common costs, and the separate costs that are not easily measured by product.*

Further, some costs are proportional to the produced volume, like raw materials and wages for personnel. Other costs are fixed and do not change with volume, like the cost for renting a factory building. Obviously, as production increases, at some point it becomes necessary to build or rent a bigger plant. This is however a stepwise change in the fixed costs and not a variable cost.

Variable cost. *A variable cost depends on the production volume of the product.*

Fixed cost. *Fixed costs are independent of production volume.*

Costs can be increased by a variety of drivers. These cost drivers can be used for modeling and calculating the costs for producing a given product or service. Generally there are two types of cost drivers that can be identified: *volume related* and *structure related*. Examples of typical cost drivers used in calculations are given in Table 3-1. The volume related cost drivers follow the volume of activities, while structure related cost drivers increase with the complexity of the venture and costs invoked by overhead work to cope with this complexity. Traditionally the focus for measuring costs has been on volume related cost drivers, however the structure related cost drivers are for example used in known methods like activity based cost analysis [Ljung et al 1994].

Table 3-1 Examples of Cost drivers

Volume related cost drivers	Structure related cost drivers
Number of produced items	Number of articles
Direct labour hours	Number of components
Wages level	Variety of components
Value of used raw materials	Number of customers
Sold pieces	Number of suppliers

The cost driver model points out issues where cost can be reduced. The volume related cost drivers show traditional areas of cutting back costs and the structure related cost drivers points out that there are a number of other areas where good savings can be made. Depending on the business model of a specific company and the structure of a specific industry, different costs become more important. The *production volume* and the *length of the product lifecycle* are important factors in the analysis of costs [Ljung et al 1994]. In a mass-producing industry with large production volumes, like the automotive industry, the variable costs of the product become important as small savings on each product makes large savings in total. Thus, with large production volumes the fixed become less important, however the fixed costs should not be forgotten. The benefits of large production volumes are generally referred to as *economy of scale*. Further, for products with a long lifespan that require maintenance, like cars and trucks, it is possible to make large cost reductions by sticking to few standardized components, reducing costs for maintaining stocks of spare parts.

With a background on costs it is possible to define cost-efficiency. It can be pointed out that cost analysis only considers the costs necessary to meet a requirement. Possible revenues collected due to performance improvements and other improved values are not considered, these aspects are however important for the cost-efficiency of a product. The definition of cost-efficiency used in this thesis is given in equation (1).

$$[Cost\text{-}efficiency] = \frac{[accumulated\ product\ value]}{[overall\ costs]} \tag{1}$$

It is thus possible to improve cost-efficiency either by reducing costs associated with a given product or by improving the value of the product without incurring a higher cost. Any new solution must be cost-efficient as there is no inherent value in changing the implementation technology. If the functionality and performance of the product is similar to the prior solution the new solution must be cheaper or it should not be introduced.

A problematic variable in this definition is the accumulated product value. The future market value of a product is difficult to estimate but at the same time very important for decisions regarding cost-efficiency in the design process. It is also difficult to evaluate the value of some soft attributes like impact on the environment and influence on third parties, such as safety measures for pedestrians. Balancing improved dependability that increases costs is a delicate business problem that contains many unknown variables and also includes ethical considerations; the ethics will not be further developed here. Further discussions on cost-efficiency will focus more on how dependability requirements drive cost and relate less to evaluating the increased value of the product.

3.1.1 Costs throughout the product lifecycle

It is possible to label four different types of costs related to vehicles: development costs, production costs, maintenance costs and availability costs. Each of these costs address different phases in the product lifecycle and is given by different system properties. All of the above introduced costs are expected to eventually propagate back to the OEM in terms of market value of the vehicle; a vehicle that is expensive to maintain will have to be priced lower in order to keep the customers. This propagation can be *direct*, regulated through agreements between OEM and customer, or *indirect* regulated by the market demand and pricing in the long run.

Development costs

Development costs are structure related and can increase due to overly complex components in the system, but development costs can also increase due to the opposite situation with a large number of simple components in the system. Either a lot of effort is spent on developing specific components or the effort is spent on integrating components.

An important issue relating to development costs is the expected sold volume of the designed component and the possibility to reuse the designs. Development costs are a fixed cost and thus the impact of high development costs increases if the production volume of the product decreases. Obviously, if very few components share high costs the cost-efficiency will be highly affected. If some parts of the design can be reused, either in a different parallel component or in the next generation of the existing component, it becomes possible to achieve a better cost-efficiency. Thus, a high development cost for a single component can be acceptable if the results can be transferred to other components, the separate cost of the initial component is converted to a common cost carried by several components.

Production costs

Production costs are generally immediately derived from the volume related cost drivers, such as the number of components and manual labor necessary in the assembly of the final product, but there are also other structure related factors that influence the production costs. Even if the set of components are known and optimized for cost, the price of a given component can vary during a product's lifecycle and change the foundation of the cost calculation. Some components are dropping in price while others may increase. The expected development of component prices is important if the lifecycle of the product is long.

The number of unique components, which is a structure related cost driver, also influences the production costs. With a standardized set of modular components it is possible to improve procurement and logistics and reduce these costs.

Further, as already mentioned, the time needed for assembly is also an important factor. A design that looks neat on the drawing board and includes a cheap set of components may still be very expensive if a lot of manual labor is required in the assembly of the complete system. It is important to be aware of the labor intensive stages in assembly. For an electrical installation many connections may be much more costly than long cables if the connections are performed manually in a time consuming way, and the other way around if the cables can be rapidly connected.

Maintenance costs

Costs for maintenance includes both repairs and upgrades and can be derived from a range of factors such as component prices, time required for repairs and maintenance, modularity and backward compatibility of components. The price of a given component will change over time and the development of the price is related to the availability of the component in the market. If the component is removed from a standard range and is no longer easily available the procurement costs will increase heavily, it might even be necessary to purchase a stock of items to ensure the supply of spare parts which means that capital costs are increased.

Obviously the price of individual components is important, but it is also important that components are sufficiently modularized such that the replacement of one component does not require the exchange of other components. It must be easy to find faults in the system and trace them to a specific component that is easily exchangeable; this requires good diagnostic possibilities as well as modular components. If repairs are easy the time to repair is obviously reduced.

Further, changes in the design must include support for cost efficient maintenance. The lifecycles of products are constantly decreasing which implies that systems must be more flexible for technology insertion, evolution and change. This means that added functionality should be possible to achieve with a minimum of alterations in other parts of the system. The functionality should be contained in a given system boundary that is well defined.

Related to maintaining compatibility when introducing new components and added functionality, it is also important from a logistics perspective to maintain backward compatibility between different generations of components. If components are not backward compatible the replacement of one unit may require the replacement of a range of related units, thus increasing maintenance costs.

Availability costs

If, for some reason the vehicle is disabled a cost is immediately inflicted. There are two main contributors to this cost, one is the capital cost and the other is the cost for substitution of the service. If the vehicle cannot be used the owner still has to cover the cost for the capital invested. Interests on loans must be paid also for the time during which the vehicle is not available. Also, the vehicle can be expected to be used for a purpose, fulfilling a transportation service. The alternative solution for this transportation may be very costly. The transportation may also be very time critical in which case the costs may increase even more as a very quick solution is needed. These increased costs are expected to propagate back to the OEM as described in the introduction of this section. The availability costs are heavily accentuated for commercial vehicles as they are constantly used for production of transportation services. Loss of availability immediately means loss of revenue as it is impossible to deliver the customer service.

3.1.2 Basic solutions to improve cost-efficiency

This section provides some general solutions aiming at improving cost-efficiency. Several methods to reduce costs have been developed such as; simple designs, standard components, product line architectures, diagnostic systems and modular systems.

Simple designs

Utilizing simple designs with a minimum of redundant components it is possible to achieve cost-efficient solutions with positive effects for production and maintenance costs. The production costs can obviously be reduced by reducing the number of components in the system, but also by reducing the number of articles used. Similar benefits apply for maintenance costs due to easier repairs and fewer components to keep in stock. The idea can be summarized in the heuristic: *Simplify. Simplify. Simplify* [Rechtin & Maier 1997].

Improved diagnostics

In order to reduce maintenance costs advanced diagnostics of vehicles based on electronics have been developed. Traditionally diagnostics were only available in the workshop as a support for the technicians. Today on-board diagnostics are used that provides early warnings for repairs and maintenance needs; the benefits of diagnostics have even led to legal requirements for advanced diagnosis systems. The use of improved diagnostics leads to reduced maintenance costs and also reduced availability costs as repairs can be foreseen and planned. The field of diagnostics is rapidly developing and telematic solutions for even more efficient maintenance solutions are just around the corner [Ogawa & Morozumi 2002].

Standard components

Open standards have been utilized in other industries to achieve cost-efficiency; it has been shown to be very effective in the avionics industry where the production volumes are significantly lower than in the automotive industry. It is however probable that the automotive sector could achieve similar gains through economy of scale. With open standards it is possible to achieve procurement with a range of commercial off-the-shelf products (COTS). Standard components and COTS have the benefits of increased production volumes and competition among suppliers, two factors that reduce the component cost.

It is important to find standards that are stable and technologically relevant for a long period in time in order to achieve the greatest advantage of economies of scale. It is a problem for evolving systems that standards, typically being a compromise of the contemporary technology, may become rigid as today's COTS may not fit the solutions of tomorrow. It is important to consider how compatibility is ensured both for old components in the new system, and also for new components in the old system.

Standards should be found in a collaborative manner and implementation details should be left unspecified whenever possible without compromising the compatibility. The requirements should be based on system issues and not detailed implementation issues. It may however be necessary to specify details due to legacy reasons. In order to achieve compatibility with existing systems, interfaces must sometimes be specified in detail in accordance with the current solution.

With requirements generally specified at system level, together with clear specifications of system boundaries and interfaces, a better optimization of the subsystems towards cost is possible. The optimization can be achieved as more freedom is given to the subsystem supplier to make reusable standard components. The performance and reliability of the components should be specified; the implementation details should be optional. Using good standards also allows modular components and the design of efficient product line architectures [Winter 2002].

Product line architectures and platform based development

In order to reduce development costs ways to increase the possibility to reuse development efforts are introduced. The reasoning is that stable intermediate forms, often referred to as platforms, need to be defined that can be reused among product variants and also among evolving variants. A layering of implementations is introduced that makes system boundaries clear. The layers specify collections of components in the form of platforms that becomes available for reuse.

Although good for reuse, general platforms may be too rigid to be cost-efficient, sometimes being over specified and sometimes lacking services. Based on the platforms it should be possible to easily produce a variety of products that can be tailored to the customer, this however also requires a modularization strategy for components. Introducing a product-line architecture, with a variable platform instead of a rigid general platform, is a good way to improve the cost-efficiency of complex systems and still enabling reuse of development efforts. The development costs can be distributed over a range of products and also the maintainability costs of the product improves [Thiel & Hein 2002].

Modular systems

The step beyond a product-line architecture is a completely modular product. The basic idea of modularity is to improve the possibility for variety management. Modularity allows the design of products that satisfies varying requirements through the combination of distinct building blocks. Modularity also improves the ease of product maintenance, reuse and recycling. Further, modularity has benefits in development and production. In development, modularity allows concurrent engineering. In production, modularity allows concurrent assembly where modules can be preassembled separately from the final assembly of the complete product [Blackenfelt 2001].

Modular systems have been found to be an important cost saver in the automotive industry. Using modular components reduces complexity in the supplier base as a single supplier is responsible for a specific subassembly. This means that OEMs are outsourcing some of the traditional territory of assembly to selected suppliers. Because of the increasing complexity of automotive systems, much due to evolving EE-systems, a reduction of complexity in some areas is necessary to keep the structure related cost drivers at bay. The cost of coping with low level integration is transferred to the suppliers while the OEMs can focus on system level integration. Volkswagen for example has explored the possibilities of inviting the suppliers into the factory at a plant in Brazil, making the supplier responsible even for final assembly of their components. The organizational issues of responsibilities are not to be discussed in the context of this work but the improved possibilities for integration through modularization should be recognized [Collins et al 1997].

In order to find good modules many aspects must be considered. The functional relationships within a given subset of parts are essential, but for cost-efficiency reasons it is also important to consider strategic module drivers. Identifying strategic drivers allows the parts to be categorized according to strategic variables, if the parts have similar strategic properties they should be clustered and modularized in order to achieve cost reduction. Driver variables can include categorization pairs such as make-buy, reuse-develop, commonality-variety and carry over-change. It is, for example, obviously better to bundle hardware and software that are both bought from a supplier into a module than to bundle in-house specified hardware with bought software [Larses & Blackenfelt 2003].

3.2 Basics of Dependability

It is common to casually name different system properties that are related to dependable and safe operation of a system. A set of broad definitions that are fairly well accepted have been introduced by Laprie [1992], later and parallel work have created some variation on the theme but the basic ideas are recurring. Laprie introduces the concept dependability as the most general concept of safety-critical properties. A slight rephrasing of his definition is given by Storey [1996]:

Dependability. *Dependability is a property of a system that justifies placing one's reliance on it.* [Storey 1996]

Laprie [1992] introduces six attributes of dependability: availability, reliability, safety, confidentiality, integrity and maintainability, as illustrated in Figure 3-2. Depending on the type and applications of the system, different facets of dependability can be emphasized. He further introduces security as an aggregate concept of confidentiality, integrity and availability.

Figure 3-2 Dependability attributes (based on Laprie 1992)

The issues of security in terms of **confidentiality**, defined as the *non-occurrence of unauthorized disclosure of information* [Laprie 1992] will not be covered further in this work. The integrity aspect will be implicitly covered. Lapries' definition is:

Integrity. *Non-occurrence of improper alterations of information leads to integrity.*

The attribute of **maintainability**, as the *aptitude to undergo repairs and evolutions* [Laprie 1992] is important and highly related to the cost-efficiency. It is important to note that the attribute maintainability also includes the evolutionary aspect of flexible system redesign and improvement. Maintainability is also a differentiating factor between reliability and availability. Definitions of these concepts as given by Storey [1996], close but slightly alternative to Laprie, are introduced below:

Reliability. *Reliability is the probability of a component, or system, functioning correctly over a given period of time under a given set of operating conditions.* [Storey 1996]

Availability. The availability of a system is the probability that the system will be functioning correctly at any given time. [Storey 1996]

Reliability and availability are closely related as they both relate to minimizing the lack of service. A high reliability will improve the availability, the difference lies in that the availability also contains the time necessary for a recovery while reliability only registers the conditions leading to a breakdown resulting in a need for recovery. The availability is thus also related to the maintainability in the sense of the aptitude to undergo repairs. Laprie [1992] suggests that the reliability and availability can be considered as an aggregate as the concepts are so intimately related.

One of the most central attributes of dependability is safety. Applications that are bounded by safety requirements are usually referred to as safety-critical systems. Safety is ensured through different methods of hazard analysis and implementation of systems that counteract these hazards, hazard analysis will be further elaborated below. There have been many attempts to define safety and one definition from Leveson [1995] is given below:

Safety. Safety is freedom from accidents or losses. [Leveson 1995]

This is an absolute definition and it may be argued that safety should be defined as *acceptable loss* and then safety would be a criterion that is possible to meet. However, in order to avoid a difficult discussion on how to define *acceptable* the chosen definition represents the ideal situation. An application specific specification will determine the required degree of safety by defining the *acceptable* level of accidents and losses. [Leveson 1995]

For further reasoning it is useful to make a distinction of safety based on the cause of accidents and losses. In this report a distinction between internal and external safety will be used according to the following definitions.

Internal safety. Internal safety is freedom from accidents or losses caused by system failure.

External safety. External safety is freedom from accidents or losses caused by events outside the system specification or due to a faulty or incorrect specification.

Acknowledging these attributes of dependability makes it possible to work proactively for improved system dependability. Choosing and endorsing specific dependability attributes for a given system is an integral part of system design. The chosen subset of dependability attributes to focus on varies with the requirements of the system. Any system obviously require some level of availability but a specific system may have requirements that emphasize a subset of the dependability attributes; a database with personal information has confidentiality issues, while a news and information application has very high requirements on availability. For embedded control systems, like automotive systems, safety is often an important concern. The choice of dependability components is important for the quality of a product.

If measurable properties are chosen it will also become possible to perform benchmarking and compare different systems with each other.

3.2.1 Impairments

If (or when) things go wrong, a terminology for system impairment is useful. Three stages of problems are defined, a fault, an error and a failure. A system failure is the final and worst stage of the three and is defined accordingly:

Failure. A system failure occurs when the system fails to perform its required function. (Storey 1996)

Before a failure, or loss of functionality, occurs there are other stages of problems. System failures are a subset of errors.

Error. An error is a deviation from the required operation of the system or subsystem. (Storey 1996)

If the deviation from the required operation inhibits the system from performing its required function the error leads to a failure. Deviations that do not violate requirements are mere errors. Errors originate in faults that are dormant within the system.

Fault. A fault is a defect within the system. [Storey 1996]

Faults can occur in value, time and space [Rushby 2001]. A fault may lead to an error. If the faulty part of the system is activated the system will respond by producing an error, the fault is activated and is no longer dormant. The concepts are illustrated in Figure 3-3.

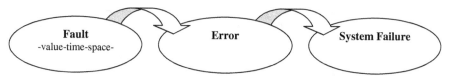

Figure 3-3 Propagation of a fault

The nature of system failures has been studied and a set of generic failure modes have been identified. These generic failure modes are used in some methods for further analysis of the cause and effects of the failure, further elaborated below in section 3.5.

These concepts of fault, error and failure apply at different levels of abstraction. A failure in a subsystem is only a fault in the higher system until the fault is activated and an error occurs. All faults are the consequence of a failure in a subsystem that has delivered, or currently is delivering, service to the observed system level of abstraction [Laprie 1992].

Consider an EMC protected memory circuit as an example to illustrate the concepts. If the EMC protection is poorly designed the protection system has a fault, if an EMC wave influences the memory to make a bit-flip an error occurs; if no mechanisms to cope with such an occurrence are built into the system, a failure, or violation of requirements, have occurred. If a corrective mechanism exists and works, as corrective coding, the error will be contained as an error and will not propagate to a failure. Further, the corrective mechanism may switch the bit-flip back and thus also remove the error. On a higher level of abstraction the memory displays a fault if a bit-flip occurs. If the defective bit is used by executing code before it is overwritten an error may occur if the bit influences the operation of the software. If the software thereby deviates from given requirements it suffers from system failure.

There is a range of terminology further describing faults in various ways, there are concepts regarding the origin or cause of a fault as well as the nature of the fault. Some of these concepts are useful for further reasoning. One distinction is between *random faults* and *systematic faults*. Random faults are associated with hardware component failures and the influence of external conditions. Systematic faults include all faults that are designed into the system, through the failure of a designer, and are also sometimes referred to as *design faults*.

The bad EMC protection in the example would be a design fault as all EMC waves with certain properties would systematically cause a bit-flip. The bit-flip on the other hand depends on the occurrence of waves and is a random fault in the memory.

Another useful distinction regarding the timing properties of faults is between *transient*, *intermittent* and *permanent* faults. Transient faults are temporary faults that originate from the physical environment. Intermittent faults are temporary faults that are the result of rarely occurring combinations of conditions, and thus are hard to reproduce. Permanent faults are always present and are easily reproduced. Systematic faults are always permanent faults [Laprie 1992].

For the purpose of settling responsibility of faults and failures the concepts of *primary*, *secondary* and *command* faults are valuable. Primary faults and failures occur within the specification of the system and are usually the result of defective design, manufacture or construction. Secondary faults and failures are the result of excessive environmental stress that exceeds specifications, while command faults are improper operation of the system that may lead to a system failure. Command faults can be created by human operators but also by failing control systems [Leveson 1995].

3.2.2 Basic strategies to deal with faults

There is a set of measures that can be used to attack faults at different stages in the system design lifecycle. Basically three different groups of techniques can be found, *fault avoidance*, *fault removal* and *fault handling*. [Storey 1996]

Fault avoidance techniques are used for prevention of faults during the design stage. Fault avoidance relies on a good design process and can be supported by formal methods. The concept includes methods to build good specifications and also to make fault free implementations of the specifications. Fault avoidance can be seen as a means for improving both the external and internal safety level of the system.

Fault removal is based on testing techniques. Fault removal is used after the design but before the system goes into actual running service. The techniques aim at making sure that the implementation is a correct interpretation of the specification, thus fault removal mainly deals with internal safety.

Fault handling is used during run-time service and is a methodology that copes with faults in order to maintain some system functionality. Fault handling is related to the internal safety of the system. The fault handling becomes very important in safety critical applications as the choice of different strategies have far reaching consequences for the system. A discussion on different fault handling strategies requires that some system descriptive terminology is defined. There are three important system properties related to the system response to faults that need to be defined, fail-silent system, fail-safe system and graceful degradation.

Fail-silent. A fail-silent system always responds to faults by omitting the output.

Fail-safe. In a fail-safe system a defined safe state exists, and it is always possible to enter this safe state in case of a fault or failure.

Graceful degradation. A system owns the graceful degradation property if a fault does not cause system failure but reduced system functionality.

The ambition with the fault handling can be set to different levels, aiming at different levels of system functionality at the occurrence of a fault. Consider a fault that could propagate to a system failure if no fault handling is applied. With full fault-tolerance no effects of the fault are seen at system level. With a target of graceful degradation the system will lose some functionality by every fault but the basic functionality will remain. Finally, the target can be fail safe shutdown, which means that the system will no longer be functional but come to a halt in a controlled and safe way if a given fault occurs. The strategies are summarized in Table 3-2.

Table 3-2 Availability ambitions and fault handling strategies

Fault handling strategy	Availability aim
Full fault tolerance	No loss in system functionality
Graceful degradation	Some loss in system functionality
Fail-safe shutdown	Total loss in system functionality

In order to achieve any degree of fault handling it is necessary to avoid single points of failure where fault handling is impossible. It is a common requirement for safety-critical systems not to allow a single point of failure

Single point of failure. A system has a single point of failure if a single fault in a component can cause a complete system failure.

3.2.3 Risk and Hazard analysis

Another set of basic concepts are related to risk and hazard analysis. Safety aims at removing threats to human life or the environment, these threats are identified as hazards.

Hazard. *A hazard is a situation in which there is actual or potential danger to people or to the environment.* [Storey 1996]

A hazard can turn into an incident or an accident. The concepts of hazard, incident and accident are comparable to the classification of fault, error and failure. The hazard is dormant until activated, and then it becomes an observable incident, similar to an error. If the incident occurs under a specific set of circumstances the hazard turns into an accident. Accidents are a subset of incidents comparable to failures being a subset of errors that inhibits the system from performing its intended function.

Incident. *An incident is an unintended event or sequence of events that does not result in loss, but, under different circumstances has the potential to do so.* [Storey 1996]

Accident. *An accident is an unintended event or sequence of events that causes death, injury, environmental or material damage.* [Storey 1996]

Associated with each hazard is a risk. A hazard analysis aim at quantifying this risk and contains an analysis of both the likelihood and the consequence of an accident. There are a variety of different definitions of the risk concept and how risk should be calculated but all measures are given by the severity and/or statistical measures on the likelihood of the outcome. Often these two factors are combined in the risk concept.

Risk. *Risk is a combination of the frequency or probability of a specified hazardous event, and its consequence.* [Storey 1996]

The value of quantifying risks is questioned [Leveson 1995], but several rather similar quantitative measures for comparison exist [Storey 1996, FMV 2001]. The risk associated with a given hazard can be calculated by giving the quadruple (p_1, p_2, E, S). Where p_1 is the probability of a hazard being activated and p_2 is the probability of the activated hazard leading to an accident. p_2 is given in probability of accident per hazard (i.e. 1/20 means that 5% of the hazards lead to an accident). The probability of a hazard can be measured similarly to reliability and availability per time unit or per situation. How the probability is quantified also depends on how the system is used, if it is continuously utilized or if it is a single mission system. The way the probability p_1 is given affects the analysis and the exposure variable E. E is the exposure of the system, measured in time or situations to match the probability measure p_1. If the probability of a hazard is $p_1 = 10^{-3}$ *per hour*, the exposure E is given in *hours*.

S is a measure of the severity of the situation and can be given in any unit depending on the system, one unit measure could for example be monetary value, another human lives. It is also possible to give the risk without the exposure. In that case the risk will be given in the style risk per exposure unit. The concepts are illustrated in Figure 3-4. The calculation of a risk is similar to any probabilistic calculation as shown by equation (2).

$$Risk = S \cdot E \cdot p_1 \cdot p_2 \qquad (2)$$

For hazards related to internal safety the probability of the accident is similar to the reliability of the system. For hazards related to external safety the probability need to be estimated based on the environment.

Figure 3-4 Risk analysis

It is possible to argue that a chain of events are necessary for an accident to take place. This means that several incidents must coincide which adds more probabilities to the calculation as illustrated in Figure 3-5.

Figure 3-5 Chain of events

In the general case the quadruple (p_1, p_2, E, S) is extended to incorporate n probabilities where n is the number of events necessary for the accident to take place. This, more general formula is shown in equation (3).

$$Risk = S \cdot E \cdot \prod_{j=1}^{n} p_j \qquad (3)$$

The derivation of such an equation may seem futile as the different factors in the equation are extremely difficult, not to say impossible, to estimate. The importance with a numeric formula is not to calculate absolute values of risk, but rather to be able to reason about alternative solutions for safety issues.

In efforts to reduce risks and improve the safety of a system it is possible to aim both at accident prevention and at damage mitigation. The accident prevention tries to reduce the probability of an accident while the damage mitigation reduces the effect of an accident. It is also possible to evaluate the probability and consequence separately and bring them together in a separate framework. This makes the contribution of the two parts more explicit and is a common approach in standards.

For automotive applications MISRA (The Motor Industry Research Association) suggest a hazard analysis framework based on the consequence on vehicle controllability. A specific consequence is mapped to a specific requirement on the probability of the hazard; the system must then be designed to meet this probability requirement [MISRA 2001]. The framework is based on five levels based of controllability as follows [MISRA 2001a]:

Uncontrollable: This relates to failures whose effects are not controllable by the vehicle occupants, and which are most likely to lead to extremely severe outcomes. The outcome cannot be influenced by human response.

Difficult to control: This relates to failures whose effects are not normally controllable by the vehicle occupants but could, under favorable circumstances, be influenced by a mature human response. They are likely to lead to very severe outcomes.

Debilitating: This relates to failures whose effects are usually controllable by a sensible human response and, whilst there is a reduction in safety margin, can usually be expected to lead to outcomes that are at worst severe.

Distracting: This relates to failures which produce operational limitations, but a normal human response will limit the outcome to no worse than minor.

Nuisance only: This relates to failures where safety is not normally considered to be affected, and where customer satisfaction is the main consideration.

Each level of criticality corresponds to requirements in a variety of subsections in the MISRA guidelines, these subsections include guidelines for specification and design, languages and compilers, testing etc.

Standards similar to MISRA for hazard analysis and other safety guidelines exist in several variants for different application areas and issued by different standardization organizations. A general safety standard is the IEC 61508, other well known standards include the US military standard MIL-STD-882D [Amberkar et al 2000], the British military standard MoD 00-54/55/56 and the avionics standard DO-178B. A common factor in safety standards is that they prescribe ways to quantify criticality and probability of hazards, and relate these measures to requirements on the design process.

The different standards have domain specific relevant measures and requirements that map probability, consequence and requirements together. One way to implement this mapping is to use hazard level matrices, as illustrated in Table 3-3. In the matrix a hazard level is determined from the probability and consequence and then hazard control resources can be allocated to improve the safety level of the system. [Leveson 1995]

Table 3-3 Hazard level matrix [Leveson 1995]

		Hazard Consequence			
		Catastrophic	Critical	Marginal	Negligible
Hazard probability	A – Frequent	I-A	II-A	III-A	IV-A
	B – Moderate	I-B	II-B	III-B	IV-B
	C – Occasional	I-C	II-C	III-C	IV-C
	D – Remote	I-D	II-D	III-D	IV-D
	E – Unlikely	I-E	II-E	III-E	IV-E
	F – Impossible	I-F	II-F	III-F	IV-F

3.2.4 Contrasting the different facets of dependability

Dependability is defined to contain both the attributes of safety and reliability. But it is not evident that a system that shows signs of faults and errors has to be unsafe; in the same sense a well functioning system does not have to be safe. Weapon systems supply good illustrations. A weapon system should destroy and kill. The fact that the system destroys and kills is a reliability property, but ensuring that the weapon system does not destroy and kill friendly forces is a safety property [Thane 1997]. If the weapon does not work, the safety criterion of not killing friendly forces is met but the reliability criterion is not met.

The definition of dependability to contain both reliability and safety is not completely unproblematic, because of the exemplified contradictions between safety and reliability. Internal safety is in line with reliability and poses no contradictions; the problem is the external safety and flawed specifications. The external safety is a factor that must be considered in the specification work in the design process. The dependability attributes of reliability and safety are graphically illustrated in Figure 3-6. In area (1) the operation is safe and reliable and in area (4) the operation is incorrect and unsafe. These areas pose no contradiction, if operation enters area (4) measures to avoid an accident must be taken, the internal safety of the system is no longer maintained.

Area (2) and area (3), on the other hand, create situations where the decision on how to take action is less obvious. In area (3) the system is unsafe even though it remains within the boundaries of the specification. The area indicates a problem with external safety. Identifying such hazards can be difficult as they are based on flawed specifications, unforeseen situations not explicitly covered by the specification or hazardous environment that cannot be changed but must be coped with. Identifying environmental hazards is an important issue and requires an environment model that makes the assumptions about the environment explicit. Adding systems that cope with the identified hazards moves situations from area (3) to area (1) improving the safety of the system. With a perfect specification of a safety critical system, area (3) would be extinct.

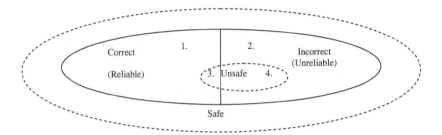

Figure 3-6 Illustration of safety-reliability inconsistencies (adapted from Thane [1997])

In area (2) the system is not well defined but does not pose any immediate hazards. This area represents a set of fail-safe modes of the system and also faults that do not threaten the safety of the system. When a fail safe shutdown is performed the system ensures that (4) is never entered by taking the system to a safe state in (2). Other approaches of fault tolerance and fault recovery moves the system from area (2) to area (1) bringing the system back into correct operation. The area outside the numbered areas (1-4) represents events outside the context of the system.

The difference and contradiction between reliability and safety is well known [Thane 1997, Leveson 1995, Storey 1996]. There lies a trade-off in that running systems sometimes need to be shut down in order to be safe. If a system is running in a hazardous state it must be decided whether to close the system down into a fail-safe mode or to keep it running, maintaining a high availability and reliability but impairing safety issues. If a nuclear power plant is on the verge of a meltdown a safe shutdown may be desirable, but at the same time the shutdown may cause disruption in the power supply at a range of hospitals and other critical institutions causing other hazardous effects. The reliability can be improved with fault-tolerance techniques while using available fail-safe modes can ensure safety. For some systems no fail-safe modes exist, in these cases fault-tolerance might be equal to safety. An obvious example is an aircraft where a system failure is obviously not fail-safe! The different concepts of fault handling are defined and further explained in the sections on fault handling strategies further treated later in this chapter.

Another issue is to consider the level of abstraction in the specification of the system. In the weapon system example of firing at friendly troops the weapon can be seen as a subsystem of a soldier, then the soldier is the malfunctioning system as he definitely is operating outside his specification, thus showing incorrect behavior and actually residing in area (4) of Figure 3-6. To cope with the safety problem the soldier system and not the weapon system should be addressed. This suggests that safety analysis requires analysis at several system levels, it is often necessary to go to the super system to establish the safety requirements.

In the work to ensure high dependability it is necessary to consider what kind of system dependability is preferred. If there is a contradiction between safety and reliability a priority decision, choosing between them, is necessary. This decision must consider the system environment and also several levels of system abstraction as exemplified in the soldier example.

3.3 Automotive requirements on cost and dependability

Cost and safety are two areas that have been pointed out as potential problem areas in the automotive industry [Feick et al 2000]. In this section the general points previously introduced are summarized and exemplified for automotive systems, and also compared to other application areas.

3.3.1 Cost sensitivity

The automotive industry is very cost sensitive because of the competitive market. Further, the automotive industry is a mass-producing industry, the economies of scale apply and the variable costs of the product are important. Anything that increases the price of an individual vehicle must be motivated by increased consumer value. The products also have a long lifespan which suggest cost reductions by using fewer variants of components and good enough reliability.

It is possible to identify some differences between cars and heavy vehicles like trucks and busses, as illustrated in Table 3-4. Trucks have a higher price, smaller series and tougher requirements on mileage lifespan. The OEMs in the heavy vehicle industry must assume a stronger responsibility for the lifetime costs of the vehicle as the customers are using the vehicles as production units and therefore will keep a stern control on the profitability of the vehicle. Availability and maintenance cost become more important, and production costs less important compared to the car segment. However, the production cost is still a core issue.

Table 3-4 Cost influences on trucks and cars

	Trucks	Cars
Production volume	Intermediate	High
Length of product lifecycle	Long	Intermediate

Another difference between cars and trucks is the supplier structure. Trucks develop more components in house and in collaboration with suppliers whereas car manufacturers are more reliant on the suppliers. This difference influences the freedom in providing solutions fit for the requirements and increases the importance of standards in the car industry compared to the heavy truck industry.

3.3.2 Selection of Dependability attribute focus

Traditionally embedded systems have been closed standalone systems which have reduced the need to consider security issues of confidentiality and integrity. To some extent this assumption is deteriorating as automotive systems become more connected to external systems, both through computer based tools in a workshop environment and also in the form of Internet applications and telematics. Intrusions exist today and may be more common in the future.

Integrity issues in terms of mistakes and unintended alterations of information should be included in the dependability obligation of the OEM. Intrusion issues are another matter, even if OEMs are seen as utterly responsible, the problem of security in computer systems is a far reaching problem outside the core competence of the automotive industry. Nevertheless, the possibility of intrusion and alteration of information must be considered for safety and reliability issues, but actually preventing unauthorized access will not be further dealt with here. Security today remains a lower priority for embedded control systems compared to information systems, but evolution in the future must be monitored for security issues.

One of the major concerns for automotive systems is safety. Vehicle systems are known to be hazardous, inflicting numerous road casualties every year. The safety issue in the automotive world has several aspects. At least three distinct *classes of safety issues* can be identified:

One reflects the *external safety* in terms of road environment and operator interface, requiring an understanding or model of the *environment* to avoid incomplete or hazardous specifications.

One concerns the *external safety* of making correct and unambiguous *specifications*; when higher order functions are introduced it is important that no unintended emergent behavior that violates the specification exists.

Last but not least are *internal safety* issues, related to system failures that may lead to incidents and accidents. Functions that are found hazardous if they fail are referred to as safety-critical functions, these safety issues are immediately related to *reliability*.

In vehicles customers also expect a high degree of reliability and availability from an economic point of view. In trucks the availability requirements are emphasized as each hour that the vehicle is not running means lack of income for the transportation company.

Further, as systems evolve it must be possible to ensure the system behavior for any combination of subcomponents. It is also important to ensure backward compatibility due to economic issues of reducing the number of articles. This is the dependability issue of maintainability that is becoming increasingly important as the number of variants of vehicle products and component articles are increasing rapidly. A way to handle the complexity by variation management is necessary.

> **Automotive Dependability Focus**
>
> **Safety** – With respect to the autonomous systems and specifications
>
> **Reliability** – With respect to actuator subsystems and basic functionality
>
> **Maintainability** – With respect to evolutionary development and backward compatibility

Based on these assumptions the further focus in this work, as illustrated in Figure 3-7, will be placed on the properties of maintainability, reliability/availability and safety, with some references to the need for integrity. Security in the sense of confidentiality will not be explicitly covered.

3.3.3 Road Safety

When safety is discussed in relation to the automotive industry it is important to define the boundaries and different aspects that are intended. Road safety is a wide concept used for the safety of the entire road traffic environment. The concept includes the safety of the driver, passengers, cargo and vehicle as well as the safety of other people on the road, the society and environment. Road safety efforts can be separated into accident prevention and damage prevention, both supported by hazard analysis [Dahlberg 2003]. The concepts are defined below and illustrated in Figure 3-8.

Accident prevention. *Accident prevention is also referred to as active safety. The purpose of the measures is to avoid accidents.*

Damage prevention. *Damage prevention is also known as passive safety and aims at reducing the consequence of an accident.*

Vehicle safety is a more narrow definition and includes the safety aspects of all stakeholders related to a specific vehicle. Road safety includes the design of road infrastructure, law enforcement and speed limits, while vehicle safety only regards safety issues traced back to a specific vehicle under given circumstances, like braking ability and crash mitigation systems.

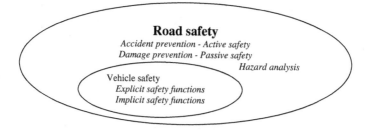

Figure 3-8 Road safety concepts

In a specific vehicle it is possible to introduce *safety functions*. *Explicit safety functions* are functions dedicated to reduce the probability of an incident or accident or to reduce the severity of the consequences. Typical explicit safety functions include airbags, safety belts and electronic stability program (ESP) systems. *Implicit safety functions* are not primarily dedicated to accident or damage prevention but they reduce the probability of incidents and support the use of explicit safety functions. An implicit safety function does not have to be primarily designed for safety but can be any function that influences the safety of the vehicle. Implicit safety functions may include adaptive cruise control, automatic climate control (relieving the driver of manual tasks) and night vision (enhancing the perception of the driver). Safety functions can be used both for internal safety, improving the reliability of the vehicle, and for external safety, coping with the hazardous road environment.

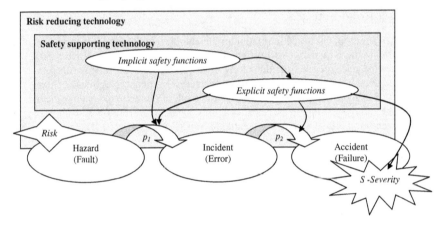

Figure 3-9 Means for vehicle safety

In the implementation of a vehicle system there are at least two things that can improve vehicle safety. *Safety supporting technology* is the necessary enabling technology for the implementation of an added safety function targeting external safety; the added functionality may include the addition of components if the function can not be covered with changes in the current implementation. *Risk reducing technology* is used for decreasing the risk due to system failures, improving the internal safety, and therefore aims at improving the reliability of the functions of the vehicle. A risk reducing technology do not add any new system functionality but may introduce new systems that supports already implemented functions for increased reliability. The terminology is illustrated in Figure 3-9.

It is important to notice the distinction between the safety supporting technology that defines the *possible functions* in the vehicle, and the risk reducing technology that improves the *reliability of functions* in the vehicle.

3.3.4 A comparison with other application areas

Cost and dependability have different impact for different applications. The requirements of manufacture-by-wire and fly-by-wire are significantly different to drive-by-wire due to some inherent properties of the systems. These properties can be described in three dimensions: The *safety criticality* or consequence of failures, the sensitivity to *system failure* (mission criticality) in operations and the *cost sensitivity* of an application, influences requirements as will be discussed in this section.

Safety Criticality

Manufacturing systems can avoid safety aspects by utilizing *fail-safe modes*. In a fail-safe mode the system is allowed to fail without any risk for damage. By this can costs from verification be reduced by accepting some downtime when new machinery is introduced, actually using the machinery in the designated process does some of the testing. When by-wire control takes the step into freely moving and safety-critical machinery the requirement on fault-free operation is significantly sharpened. There are no fail-safe modes in the avionics industry for example. This requirement increases the need for testing and verification to ensure dependability in fly-by-wire products.

In the automotive industry there are some semi fail-safe modes that can be used. However, steering and braking are safety-critical functions. It is possible for a vehicle to shut down, maintaining steering and braking until it comes to a halt, without being a danger on its own. A problem is if this happens on a heavily trafficked highway where a collision with other vehicles is probable. Vehicles on their own have fail-safe modes but they may be used in a hazardous environment that can negate the fail-safe mode.

System failure sensitivity

Sensitivity to system failures is related to the cost associated with a stop in the system. Space and avionic applications cannot afford any stops in the process because this will terminate the entire mission. Process, manufacturing and drive by-wire applications on the other hand have the possibility to be stopped and mended still acquiring a high availability even though reliability is not perfect. This possibility relaxes the requirements for reliability somewhat, but stops are still costly.

Cost sensitivity

Cost sensitivity is related to the cost added by introducing dependability increasing measures. Improving dependability is based on some kind of redundancy that brings redundant costs with it. The avionic industry with airplane and space applications are less *cost sensitive* than many other industries and it is in these industries some of the first safety-critical by-wire systems were introduced. These fly-by-wire applications use costly triple modular redundancy. The concept of fly-by-wire has been in use in the Airbus series aircraft commercially since 1983 in the A310 model [Augustine

2000]. The Airbus 320 was the first that depends entirely on by-wire control and it was certified in 1988 [Briere & Traverse 1993]. Introducing triple modular redundancy is not a feasible solution in the automotive industry.

Fault handling strategies in response to different requirements

Based on the sensitivity to dependability and cost different fault handling strategies are chosen. From one point of view there are three different levels of fault handling: Fail safe (FS), Fault tolerant (FT) and Fault detecting (FD) systems.

Fail-safe systems utilize a safe shutdown sequence to a previously known static safe state. Operability is not maintained in the occurrence of a failure. Fault detecting systems use an even simpler fault handling strategy, only alerting the operator when an error is detected, leaving the decision-making of the fault handling to other systems. These are common strategies at plants where process by-wire and manufacturing by-wire is used.

The most rigorous way to approach fault handling is to implement fault tolerance. Fault tolerance maintains operation even in the presence of faults. It may preserve the degree of service or go through a degeneration process known as graceful degradation, where a lessening degree of service is maintained. Fault tolerance is employed in avionics and is to some extent used in drive by-wire applications. The degree of fault tolerance in future automotive applications is still undecided. The issues that have been discussed and the properties of different applications are summarized in Table 3-5.

Table 3-5 Requirements by applications

Application area	Example	Cost sensitivity	System failure sensitivity	Safety-Criticality	Fault strategy
Space-by-wire	Satellites	Low	High	Low	FT
Fly-by-wire	Airbus	Medium	High	High	FT
Process-by-wire	Nuclear	Low	Medium	High	FS
Drive-by-wire	Cars	High	Medium	Medium	FS/FT
Manufacture-by-wire	Robotics	High	Medium	Low	FD/FS

3.4 Reliability

Reliability is at the core of dependability. If the system does not do what is expected from it any design based on the services of the failing system will be influenced. Reliability can be improved in a variety of ways. To begin with, reliability can be improved by using reliable components that are properly dimensioned according to the environment. Further, the implementations must be properly carried out and the assembly and testing must be sufficient in quality to maintain a high reliability. Finally, good components can be combined in a topology that achieves a system reliability that is higher than the individual reliability of the components.

3.4.1 Component reliability, robust design

As no chain is stronger than its weakest link it is evident to begin building a reliable system with reliable components, where components refer to any defined part of the system. To calculate the system reliability the reliability of the service of each component must be known. Figures for the reliability of a given component can be attained in handbooks or sometimes from the manufacturer of the component. An example of a commonly referred handbook for electronics is the US department of defense handbook MIL-HDBK-217. This handbook contains failure rates of various components based on experimental data [Storey 1996].

However, the figures of reliability are often criticized to be off the mark and also dependent on environmental factors making the statistical average reliability questionable to use. When using probabilities to calculate system reliability the results are very coarse and should be interpreted carefully. Further, the reliability of software components is hard to establish as the testing criteria for software does not correspond to a given time span. Nevertheless, it is useful to make calculations in order to compare alternative designs, either aiming for the most reliable solution or improving other qualities trying to avoid impairing the reliability [Storey 1996].

A rule of thumb says that is practically impossible to verify reliability better than 10^{-4} faults/hour through testing. The reason for this is that 10 000 hours corresponds to 417 days or slightly more than a year. For statistical verification of a component several tests must be performed. In order to achieve 95% confidence ($C=0.95$) of a reliability ($R=10^{-4}$) of 10^{-4} faults/hour a number (N) of $3*10^{4}$ hour long tests are necessary, as shown in equation (4). This number of tests is at the border of cost-efficient practical testing [FMV 2001].

$$N > \frac{\log(1 - C)}{\log(1 - R)} \quad (4)$$

One way to improve the reliability of a given component is to create a robust design. A robust design is generally achieved by over dimensioning critical attributes. For example, a load bearing beam may be designed with an increased thickness or can be given a larger dimension to achieve a more reliable and robust construction. Another example is given by an electrical connector that may be produced in a different material with better conductor properties or with better resistance towards corrosion depending on the requirements of the environment.

Fault tolerance at system level may be necessary in some cases where over dimensioning is not applicable. For safety critical applications, both in the aviation industry and in the automotive industry, the system reliability requirement is often set to 10^{-9} faults/hour. This requires that measures are taken at system level to ensure the reliability.

To ensure reliability, testing must be performed even though it is not possible to verify the system to desired reliability levels as previously mentioned. For this reason the testing must include unit tests of subsystems. These testing efforts are to some extent based on the fault hypothesis as the rate, type and number of subsystem failures (system faults) accepted in testing must comply with the fault hypothesis. Another issue is to ensure that the manufacturing and assembly of the system do not introduce any new faults.

3.4.2 Containing the faults by good implementations

Designing a good fault hypothesis is essential to ensure that a reliable system can be successfully implemented. The requirements from the expected operating environment of the system must be known and applicable for the design. A fault hypothesis recognizing the modes, maximum number and arrival rates of the faults is necessary for a reliable system design [Rushby 2001].

For a reliable implementation the effects of a fault should be minimized. In a networked embedded system sharing resources it is possible to achieve fault propagation. Traditionally, safety-critical systems have been implemented as stand-alone systems in order to improve the reliability of the system by avoiding fault propagation. Today, interconnected network are used as they can produce beneficial effects in terms of resource sharing and distributed functions. Because of this it is necessary to consider the propagation of faults in the design of communication systems and ensure *fault containment*. Fault-containment means that a fault or error that occurs locally within in a subsystem is not allowed to propagate to other subsystems. It is necessary to maintain some kind of separation boundary that contains the fault in the original region. This separation does not have to be physical, but it is also possible to have logical and temporal boundaries [Watt 2000]. Based on the basic assumptions in the fault hypothesis it is possible to extend the hypothesis with *fault containment units* (FCU). FCUs are boundaries in the system that inhibits fault propagation in accordance with a given fault hypothesis [Rushby 2001].

In software based systems FCUs are implemented through firewalls. In order to introduce firewalls the system must be partitioned in a proper way. Firewalls can separate parts of the system physically, logically or temporally. The separated regions are interconnected through safety ports that validate the communicated data. Each of the types can contain faults that occur separately in the different dimensions of value time and space [Watt 2000].

The integrity of a given system can be given in levels where the levels refer to the degree of how well improper alterations of information are avoided. Systems processing data can be associated with an integrity level and the data itself can also have an integrity level. A common way to ensure the integrity of the system is the Biba policy that prohibits flow of data with a given level of integrity into processes with a higher level of integrity. Further, data inherits the level of integrity from the processing system if the system integrity is lower than the data integrity. This creates a system with very little flexibility as information flows become unidirectional between systems of varying levels of criticality. An alternative policy is the Clark and Wilson policy that focuses on the data modification instead of the data flow. Data and processes are validated to a given integrity level. The integrity level of data can be kept only if the manipulating process has the same level of integrity. But it is also possible to revalidate data to achieve a higher integrity level. If multiple levels are to be handled without losing the flexibility there must be a mechanism supporting this validation of objects [Totel, Blanquart, Deswarte & Powell 1998]. One such mechanism was implemented and formally verified in the project GUARDS [Fantechi, Gnesi & Semini 1999].

One concept that may improve the possibility to implement fault-containment is fail-silent systems. A fail-silent unit ceases to function by shutting down. This means that any internal fault is contained within the unit and the fault will not propagate. On the other hand this strategy might unnecessarily close down a functional unit if not implemented properly. Also, the omission of service may provide a fault at a higher level in the system and thereby the fault is propagated.

3.4.3 Fault handling strategies and redundancy

All fault handling is based on redundancy, either to monitor the system or to implement fault tolerance. Redundancy for fault tolerance uses the idea that elements of the system are duplicated maintaining system functionality in case some of the duplicates fail. The redundancy can be created in many forms, each serving different purposes, tolerating different types of faults. It is possible to use redundancy in hardware, software, information and time. Hardware redundancy usually implies modular redundancy of components; software redundancy usually adds some diversity to the modules. Information redundancy implies redundant communication as exemplified by fault correcting coding. Temporal redundancy implies, for example, retransmission of messages or multiple consecutive calculations of a given algorithm [Storey 1996].

Fault tolerance can be based on passive methods that simply mask the faults by the structure of the system, or active methods that use control logic to handle detected faults by reconfiguration. The terminology active or passive redundancy is thus used to distinguish between reconfiguring systems and fault-masking systems [Storey 1996].

Fault detection should be utilized by all fault tolerance techniques, even though it is not necessary for the actual fault tolerance if passive fault masking systems are used. All active fault tolerance techniques require that the fault is detected before it is tolerated. Passive techniques can be designed without fault detection but this is a highly hazardous approach as the current degree of redundancy is unknown. A triple redundant system can have a faulty component which means that the fault tolerant ability is lost. A passive fault tolerant system should also use some kind of warning system when a fault is detected [Storey 1996].

An alternative to fault tolerance is fail-safe shutdown. If a fault is detected and a fail-safe mode exists it is possible to enter this fail-safe mode. This makes a safe shutdown, but the availability of the system is crippled. However, it may still be possible to make a recovery from the fail-safe mode to the desired running mode of the system.

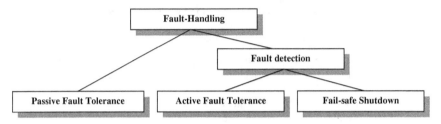

Figure 3-10 Fault handling strategies

All of these methods are based on redundancy but the active fault tolerance tries to use intelligent logic in order to minimize the need for redundancy. The active fault handling process is divided into a set of sub-steps. The active methods include detection of faults, fault-diagnosis and treatment of faults [Feick, Pandit, Zimmer & Uhler 2000]. The methods are graphically illustrated in Figure 3-10.

Fault detection techniques and fault models

Faults can be of several types. When designing a fault-tolerant system it is necessary to use a fault hypothesis as a reference for construction and validation as the fault detection is based on this fault hypothesis or a fault model. As previously stated a fault hypothesis must describe the modes (types) of faults to be tolerated, together with their maximum number and arrival rate [Rushby 2001]. A limitation in the fault detection is that faults that are to be handled must be explicitly stated in the fault model. The faults are then signaled to the system, this function should also be contained in the diagnosis function of the system.

The fault detection interprets inputs from the system according to the fault hypothesis. In the design of the interpretation it is necessary to ensure that the fault detection rate of the interpretation is in line with the maximum arrival rate of the fault hypothesis. Further it is also necessary to consider the possibility of false alarms. An accepted rate of false alarms must be decided and the design of fault handling actions must acknowledge the possibility of false alarms [Rushby 2001].

Information redundancy and fault correcting coding

Fault correcting coding is an efficient way to cope with expected faults in data and information. These codes are a way of implementing information redundancy. It is possible both to detect and correct faults through coding. If mere fault detection is used in the form of a checksum or similar measure the fault can only be corrected through a retransmission, a so called automatic repeat request (ARQ). By utilizing ARQ, temporal redundancy is used in combination with information redundancy. But it is also possible to use coding to make a correction at the receiving end, a so called forward error correction (FEC). Utilizing an FEC strategy requires more control bits and thus implies a higher information redundancy. However, the FEC does not require a return channel for retransmissions and thus is it possible to achieve fault tolerance even on a one way communication channel. Also, the absence of retransmissions avoids congestion of the communication media in case of a burst of faulty messages [Öberg 1998].

There are essentially two kinds of fault correcting coding. Either the bit-stream is divided into blocks that are encoded and decoded separately, so called block coding; or the code is applied continuously on the bit-stream, so called convolution code. The coding methods are illustrated in Figure 3-11.

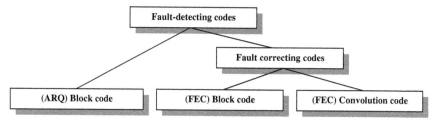

Figure 3-11 Fault handling codes

For a given code it is possible to define the number of faults it can correct, the number of faults it can detect and the overhead introduced. A given code has a Hamming distance that can be calculated. Hamming distance is a commonly used property of fault correcting code. If the Hamming distance d_{min} is known the number of faults, t that the code can correct is given by equation (5).

$$t = \left\lfloor \frac{d_{min} - 1}{2} \right\rfloor \qquad (5)$$

A commonly used method and well known acronym related to coding is Cyclic Redundancy Check (CRC). This is the label for cyclic block codes that are used for fault detection only [Öberg 1998].

Temporal redundancy by retransmission and recalculation

If transient faults can be expected it might be useful to have some temporal redundancy in the system. One way to implement temporal redundancy is to transmit messages on communication buses several times; a similar approach for calculations is to recalculate a value several consecutive times. For each of these methods a single transient fault can be masked or detected by utilizing a voter comparing the results of the transmissions or calculations. A problem with the method is that it will not be very tolerant to bursts of faults, as the recalculations and transmissions are assigned closely in time. It is not uncommon that faults occur in a bursty manner thus reducing the value of temporal redundancy. However, temporal redundancy is a useful method for applications that are rarely affected by fault bursts.

Passive fault tolerance by modular redundancy

Passive fault tolerant systems do not need to utilize fault detection; instead faults are masked to achieve fault tolerance. However, the importance of using fault detection even if it is not necessary for the fault tolerance functionality is commonly recognized. It is important to know if there are errors present that need service and repairing. Some of the passive fault tolerance techniques are integrated with active techniques, creating hybrid techniques [Storey 1996].

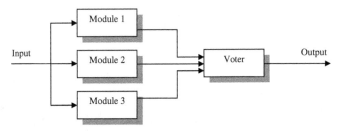

Figure 3-12 Simple TMR arrangement

A common technique, heavily utilized on hardware in the avionics industry, is the triple modular redundancy (TMR) technique. In a TMR system a component is triplicated to avoid system failure if one component fails. The three modules receive identical input and a voter compares the outputs of the modules. If one of the modules output differs from the other two this output will be discarded by the voter. A simple TMR system is shown in Figure 3-12.

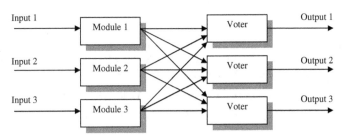

Figure 3-13 Multiple input/output TMR arrangement

It is still possible to have a single point of failure on the input or output if these are not triplicated too. Figure 3-13 shows a TMR arrangement with multiple inputs and outputs.The triple modular redundancy technique can obviously be extended to a higher number of modules. This is often referred to as N-modular redundancy (NMR). Introducing higher redundancy calls for more complex voting algorithms to find a proper output [Storey 1996].

Diversity in the redundancy

Using simple duplication of components in the system offers some protection against random faults, but is useless against systematic faults as the identical units will generate similar faulty output. To deal with this problem, redundancy can be combined with the choice of diverse components. Diversity is the only way to build fault tolerant systems that tolerate software faults as software only contains systematic design errors and no transient errors. This means that redundancy in itself is meaningless for software purposes [Storey 1996].

It is possible to distinguish between three types of redundancy based on the degree of diversity: *replication, functional* and *analytic* redundancy.

Replication *is a mere duplication of a system.*

Functionally redundant *systems produce identical output based on the same input, but allow internal diversity.*

Analytically redundant *systems are allowed to produce different outputs if all the results comply with a defined model, the systems are said to show well formed diversity.*

An example of analytical redundancy is the pressurized cabin and the oxygen mask in an aircraft. Both systems comply with the function of supplying oxygen to the passengers. Analytical redundancy can be used for graceful degradation of a system [Sha et al 1996].

Diversity introduces a delicate problem for voters. In a general system voting can be performed either in hardware or in software and there are many ways to decide on the output. An important distinction is between *exact* and *proximate* agreement. Proximate agreement uses separate diverse systems that find separate values that are compared for validation in the end using some kind of threshold or deviation algorithm. The idea is to avoid common mode failures by using diversity in the calculations. A problem with the approach is that small differences in the early stages may create large differences in the end and it is difficult to actually know which result to trust. Further, the choice of thresholds and integration of values become complex and prone to faults [Lala & Harper 1994]. In an investigation of flight tests documented by Mackall [1988], as referenced in Rushby [2001], all observed failures were due to bugs in the design of the fault tolerance mechanisms themselves which suggest that proximate agreement may be problematic.

With exact agreement calculations are performed one stage at a time with identical inputs and the results are compared and required to be identical. This requires that data can be distributed consistently throughout the system, a property known as *interactive consistency* or *atomic broadcast* [Rushby 2001].

Active fault tolerance techniques

Using passive fault tolerance to mask faults can be expensive as a very high level of redundancy is necessary. Three units are necessary to mask a single fault by majority vote and five units to mask two faults in the same manner. An alternative is to use active fault tolerance that uses fault detection on a given system, and in case of a failure switches to a *standby system*. Each standby system can cater for one faulty unit. Standby units can either be arranged for hot standby, with the spare running in parallel with the original system, or cold standby where the spare needs to go through a startup routine before switching can take place [Storey 1996].

An alternative arrangement is to utilize *self checking pairs*. In this design two replicated modules are fed with the same input and the outputs are compared. If the results from the two modules do not match, a signal indicating failure is activated. If the failure indication is inactive, a correct output is expected and the output from one of the modules is fed forward [Storey 1996].

When designing an active fault tolerant system there are a set of parameters that must be decided. The fault model must be decided with the interpretation of a fault and also the number of tolerated faults. In the fault handling strategy the speed of reconfiguration and degree of degradation in case of a fault must be considered and also the possibility to reintegrate faulty units [Rushby 2001]. Another issue with active fault tolerance is that spare subsystems must also be checked for faults [Avizienis 1997].

If active fault tolerance is utilized it is possible to choose diverse alternative subsystems as backup. These subsystems may have a reduced functionality which means that a graceful degradation will be achieved if the main system fails. The failed systems may then recover, either through backward recovery, returning to a previous error free state or through forward recovery, constructing a new valid error free state. A recovery procedure generally includes fault diagnosis and removal, error elimination, state restoration and recovery validation [Avizienis 1997].

3.4.4 Cost-efficient Fault Tolerance

In order to replace critical functions in the vehicle and to adhere to current legislation that demands a redundant braking system, some fault-tolerance is unavoidable. The only way to implement fault-tolerance is through redundancy, and adding redundancy is a very costly thing to do. The introduction of redundancy must be well motivated before the added cost becomes cost-efficient. The safety, reliability and cost aspects are approached from different angles by different researchers.

The inclusion of mechanisms that tolerates single points of error within the system with maintained critical functionality is consensus. This level of fault-tolerance is seen as necessary for the by-wire system to exceed current systems in the field of safety [Sallee 2001, X-by-wire 1998, Scobie et al 2000, Sanfridsson et al 2000]. An interesting point to make is that it is not clearly defined what a single point of error is. It is necessary to define at what level faults and errors should be considered. Is an error defined as an error in the entire braking system, or an error in one of the brake actuators, or an error in a part of the actuator? Building fault tolerance requires a fault model that describes the errors that you require to tolerate.

The controversy of fault-tolerance is if it should be applied *systematically* or *application specific*. Researchers in the field of safety and fault-tolerant computing suggest that fault-tolerance in general is needed for complex systems to survive, and systematic fault tolerance is preferable [Avizienis 1997, X-by-wire 1998]. Lala and Harper [1994] recognize and welcome a progression from ad hoc techniques to more systematic design of safety critical systems. Others advocate application specific fault tolerance arguing for a more cost-efficient solution without loss in fault tolerance [Feick et al 2000, Scobie et al 2000]. The application specific fault tolerance utilizes less redundancy but must be carefully redesigned if components in the system are changed or reused for other applications.

Application specific fault tolerance

The main argument for application specific fault tolerance is cost efficiency. In an application specific fault-tolerance implementation the characteristics of the system is used in decisions regarding the level of fault tolerance for specific subsystems. Through this technique unnecessary redundancy is avoided. But, this also means that one subsystem becomes dependent on other subsystems to guarantee system level fault tolerance. Thus, changes in one subsystem can have far reaching consequences for the entire system behavior and the maintainability of the system is reduced.

The end-to-end argument also supports application specific fault tolerance. The end-to-end argument was initially applied to computer communication and file transfers. The argument says that a check of the end-to-end file transfer is always implemented, thus the end to end performance and reliability is all that counts in the design, no internal checks are necessary for correctness [Saltzer et al 1984].

Further, the reliability of a communication system should only be increased as long as the communication performance increases. Assuming that a failed transfer requires a retransmission, initial efforts to improve reliability will improve the performance such that messages do not have to be retransmitted. However, at a certain level the overhead introduced from reliability measures will impair the actual ability to communicate. The trade-off concerns performance and is not related to a requirement on correctness [Saltzer et al 1984]. It is also possible to assign different subsystems different levels of fault-tolerance through application layering. Then, the end-to-end argument can be used for each layer. The interfaces to a given layer should have a given performance and reliability, but the internal parts of the layer can be designed freely.

If application specific fault tolerance is used it is important to use redundant system characteristics that cannot be changed or are not easily changed. It is also important to be aware of the dependencies for fault tolerance and to document these dependencies thoroughly. Scobie et al [2000] suggest a cost efficient brake-by-wire architecture using both the (unchangeable) inherent system redundancy of four wheels and an application layering strategy.

Systematic fault tolerance

The problem associated with proximate agreement in analytically redundant systems is one argument for implementing systematic fault tolerance [Rushby 2001]. Systematic fault tolerance is achieved by system partitioning, followed by fault tolerant subsystem design and system-wide integration. In the partitioning step building blocks and communication links are defined to identify replaceable error containment boundaries. These blocks, creating subsystems, can and should be duplicated in order to achieve systematic fault tolerance [Avizienis 1997].

Within subsystems, errors need to be detected and a recovery strategy needs to be formulated in order to achieve fault tolerant behavior. For systematic fault tolerance a strategy for a coordinated system-wide integration is also necessary. In a systematically designed system each block of subsystems should be independent of the behavior of other subsystems [Avizienis 1997].

The application specific approach can be more efficient and quick if the system works in a well known environment and is not supposed to undergo any changes in the future. The systematic fault tolerance is more tolerant to new faults introduced by system or environment changes. However, no matter what approach is used the performance always relies on the quality of the fault hypothesis employed. The fault hypothesis must contain all the faults the system is designed to tolerate.

3.4.5 Reliability and Costs

Improving reliability is always costly, however for any solution that improves reliability the availability costs decrease and the maintenance cost may also decrease if fewer repairs are necessary. The value of the benefits must be weighted against the increased production cost based on the properties of the specific application.

One solution is to use more reliable components; this generally means more expensive components or simpler components with less functionality. For example, two ways to solve the problem with unreliable connectors is to either use gold plating of the connector, or to introduce more complex mechatronic modules where the connections can be removed. Actually, both solutions increase the component costs of the system but the latter solution moves the problem from the connector to the mechatronic module. With increased component costs the production and maintenance costs for each repair are increased.

If more reliable components cannot achieve the targeted reliability redundancy must be utilized. Redundancy can be added in the dimensions of physical hardware, software, information and time. For each dimension the level of redundancy must be chosen. With redundancy there are components and cost added to the system without adding any more functionality while the system is up and running, this means that the costs inflicted must be saved by the reduction of availability costs.

Redundancy can be applied in either an application specific or a systematic fashion. An application specific approach targets the parts where the most "bang for the buck" is achieved, the systematic approach is more elaborate and thorough.

Table 3-6 Reliability methods and costs

Reliability Method	Cost type			
	Development	Production	Maintenance	Availability
Reliable components	0	☹	☹/☺	☺
Application specific redundancy	☹	☹	☹☹	☺
Systematic redundancy	☹	☹☹	☹☹	☺☺
Diversity	☹☹	☹	☹	☺

With redundancy comes the possibility to use diversity. Diversity may further improve the reliability and availability but it introduces even more costs in production and maintenance as procurement of a higher number of alternative components is necessary. The variants must be available as spare parts and also maintained in the development organization. Further, diversity doubles the development effort as the different components should be developed and maintained separately. The implications for costs related to reliability are summarized in Table 3-6.

3.5 Safety

Safety is pointed out as one key element of dependability that automotive systems must care for. This section expands some of the issues with safety by discussing the threats to a system based on the separation of internal and external safety. Further, a range of methods and technologies that can be used to cope with the threats are discussed. In the end the methods are evaluated from a cost perspective.

The difficulty for safety analysis is that some of the threats are unknown. A hazard analysis may give a range of situations and conditions that the system being designed must be able to care for in order to be safe, but there is always the possibility of things going wrong, either from unforeseen situations or because of unexpected misbehaviors of the system. The system can be built with wrong or incomplete models of the system environment and operators in mind. Further, functions within the system may interact in a way that was not expected or accounted for, the specification of the system is wrong or incomplete. Also, hazards may arise due to malfunctioning components.

Extending the distinction of external and internal safety, it is possible to find two classes of hazards, internal and external hazards. In line with safety the following definition is proposed:

Internal hazard. *An internal hazard is a fault that can lead to a system failure in which there is actual or potential danger to people or to the environment*

External hazard. *An external hazard is a situation, caused by events outside the system specification or due to a faulty or incorrect specification, in which there is actual or potential danger to people or to the environment*

3.5.1 External hazards in the environment

Anything around a system can be a hazard. This makes general conclusions regarding external safety very difficult, the space of possible situations is infinite. One problem is to define the borders of responsibility of a system. For an automotive system it is occasionally discussed whether the OEM or the driver is responsible for a given accident or damage incurred. Obviously, it is impossible for OEMs to take full responsibility for a vehicle, as situations may arise that are outside their field of control.

The infinity of external hazards makes it important to clearly define, for development and litigation purposes, what is within the responsibilities of the system and what is not. Larses and Chen (2003) propose that the environment should be modeled explicitly with environment models for this purpose. The models should show the situations where the OEM assumes responsibility.

In general, external hazard identification is performed through a set of ad hoc methods. Some methods for preliminary hazard analysis exist, like brainstorming sessions, but much of the actual work is performed tacitly by experienced engineers, more in the sense of an art than in the sense of engineering.

3.5.2 System Safety and the Operator

Another design issue of the embedded control systems the interaction with the operator. For vehicular applications there are several options available both in terms of man-machine interfaces and human-machine control boundaries.

The operator can be seen as a part of the system or as a user of the system. Also, it is possible to see the operator as a support to the autonomous system instead of the traditional way of regarding the system as a tool supporting the operator in a task. The border between operator control and autonomous behavior is here labeled the control boundary. The design of operator interfaces is closely related to the choice of control boundaries. The interface grants access to a specific decision space of given dimensions defining a coarse control boundary, the control boundaries also specifies how much of each dimension that can be used by the operator.

Human-Machine interfaces

The interface between operator and machine has been the target for prior research for example by Modugno et al [1996]. Two properties have been identified as important for safety. The human-machine interface needs to *lack ambiguity* and be *robust* in the design so it is impossible for the operator to create an unstable situation.

When choosing the interface to the control system it is an essential consideration to decide to what extent the properties of the operator are to be allowed to influence the system. The choice of interface limits the control of the operator to a specific *decision space* that can be reached through the available interfaces

Within the scope of the interface design it is important to consider what type of control the operator is to impose on the system. The operator can either be in control of actual *actuators* or given *system parameters*. The system parameters can be at different levels of abstraction that allows different levels of direct control. For example, each new actuator or system parameter placed under the control of the operator adds a dimension to the decision space.

Introducing more automated control creates control environments more prone to be based on abstract system parameters. An example of such development is the transmission of a car. The original transmission uses a clutch and a gearshift stick allowing the driver control both of the choice of gear and how the gears are shifted. The control is very much in the hands of the drivers' skill and personal driving style. With a modern automated gearbox, where the clutch can be removed, it is possible to choose between fully automated transmission and automated gearshift where the driver chooses which gear to use at a certain point in time but the system does the actual shifting. Now, the driver has less influence over the vehicle and the skill and personal driving style becomes less important. The dimension of the decision space of the operator has been reduced due to the removal of the clutch control.

The autonomy trade-off

Making the decision space clear to the operator is very important for safety issues. *Incomplete knowledge of automation is a highly contributing factor to accidents* for safety critical systems. The factor has been identified by studies of aircraft disasters and the Three Mile Island accident [Modugno et al 1996]. Choosing the dimensions within which the operator may influence the machine is not all there is to consider in automated control. It is also necessary to specify the control range in each dimension. A safety-critical factor identified by Modugno et al [1996] is the lack of robustness of the computer when faced with an imperfect human operator. This implies that boundaries as well as sub-spaces with control inputs creating unstable situations need to be placed outside the reach of the human operator.

This issue is here referred to as the autonomy tradeoff. This tradeoff requires strategic decisions on when to let the driver be in control and when to let the system be in control. In the avionics industry this tradeoff is well known. By introducing limits on the operator control of the system it is possible to avoid failures due to pilot mistakes but at the same time new causes for accidents are introduced. Certain modes can be unintentionally entered and correct pilot responses are then overridden by the system. One such accident occurred on 24 April 1994 when an Airbus 300-600 entered an unintended mode during landing. The pilots and the control system engaged in a struggle for control over the airplane causing the aircraft to eventually pitch up to near vertical, stalling and crashing – killing 264 passengers and crew. [Olson 2001].

Another example of the problem with automation is the crash of an early version of the fighter aircraft SAAB Gripen in central Stockholm 1993 during an air display show [Neumann 1993]. During the show the pilot experienced an oscillating movement in the aircraft. In his attempts to stabilize the aircraft, he gradually increased his control movements until he finally was rapidly moving the control stick between the extreme positions available. Simultaneously, stabilization efforts were also performed by the computer system. Still reading the pilots input, the aircraft complied with the pilots commands and oscillated heavily before finally coming to a stall. As the altitude of the flight was very low the pilot at this moment ejected himself from the aircraft. After the crash the control system was redesigned by slightly reducing the automation, extending the operator control boundary to match the pilots training [Fredriksson 1993].

Different choices in the autonomy trade-off introduce different training needs for operators, the understanding of machine response to different control inputs is essential. A business issue related to the trade-off is the required flexibility for experienced drivers (with a wider control boundary) and the decreased training necessary for new drivers (with a more narrow control boundary).

A problem when determining hazardous input subspaces for vehicles is that the operating conditions of the vehicle are never fully known. At one moment a turn at high speed is necessary to avoid crashing into a mountain wall, while at another instant a lane shift at lower speed is extremely hazardous due to slippery road conditions. These two situations may be perceived similar by the control system, but the required reaction to avoid a hazard is very different. The ambiguous conditions suggest that much of the control should be left to the discretion of the driver, as the hazards of vehicle control are highly dependant on the operating conditions. But even if some of the core functions of a vehicle should remain within the control boundary, there are other issues where automated computer control can be of good assistance to the driver.

The intent of the driver cannot be questioned for design issues. It must be presumed that the driver does not want to crash with the controlled vehicle. The next consideration is to question the skill and judgment of the driver. In society, it is generally more accepted with human induced errors before technical system errors. It is from a corporate image perspective more discrediting for a company to admit technical malfunctions than to show a misuse of the equipment. This attitude can create a preference for trusting the driver with more responsibility. On the other hand are most accidents operator related, which suggest that the automated machine should take more responsibility for operations.

Often, a safe and dependable behavior of a vehicle is defined as following the intent or commands of the operator. Using the word intent would be a better choice considering the discussion above. The intent of the driver cannot be questioned in safety work, but his skill and the commands he issues can and should be questioned.

Mode confusions

Another known related problem is the issue of mode confusion [Rushby 2002]. Mode confusion occurs when a machine behaves in an unexpected way that is not anticipated by the operator. This is common particularly in aircraft where advanced automation has introduced a range of automated modes in the system. The pilot creates a 'mental model' of the system which is an expectation on the behavior, and also necessary responses, for a certain mode of operation. However, if the pilot believes that the aircraft is in one mode while the system actually is in another mode hazards arise.

An example scenario is described by Rushby [2002], originally reported by Palmer [1995]. In the example a pilot believed that the aircraft was in a mode for automatic climb to a given altitude. However, this mode had been overridden when the pilot had accelerated the climb speed manually. The result was that the aircraft missed the designated 5000 ft and climbed to 5500 ft before the crew managed to stop the climbing. The situation is colloquially called a 'bust' and can be hazardous, especially around airfields as several aircrafts are sharing a small airspace.

3.5.3 Hazards derived from system properties

Besides mode confusions between operator and system it is also possible that the system itself becomes inconsistent in its behavior. This behavior can emerge if the system complexity becomes too high and verification and validation of the system is impossible. The complexity of automotive electronics has increased as architectures has evolved from simple control units to the networks of today where distributed pieces of code contribute to higher level functions. The main trend today is to move the software to central computing units, reducing the number of hardware components but at the same time increasing the software complexity [Würtenberger 2002].

With complex functions in the system, the specifications of both functions and implementation components will have many relationships and interdependencies. The system behavior becomes hard to overview and restrict to safe behavior. The outcomes on the system behavior due to changes in a function are impossible to foresee intuitively by engineers if the complexity is too high. Due to this, tools and formalizations that maintain traceability of changes are necessary to keep the safety at an acceptable level.

Further, throughout implementation new requirements are implicitly introduced, and in addition several assumptions are hidden in the implementation. These requirements and assumptions must be documented and echoed to all the related functions and implementations during the design process. Besides adding details not covered by the specification it is also possible that faults are introduced during the implementation that breaks the compliance to specifications. It is usually non-trivial or even impossible to implement a specification without adding design faults, for software the problem is aggravated as a given design (source code) can result in different executables depending on the compiler. To cope with these problems a rigorous design and implementation process is important; a possible solution to improve the quality of implementations is to utilize code generation. Another tool supporting dependable software design is a handbook is available from MISRA with guidelines on how to use a safe subset of C in automotive applications [MISRA 2001a]. Compliance to the MISRA standard should minimize the possibility of errors induced by bad programming practice.

Design faults in the software and the possibility of electrical components breaking down adds internal hazards. Internal hazards are derived from non-reliable safety critical components. Internal hazards can be remedied by improving the reliability of the system. A purpose with the hazard analysis is to find internal hazards and components that require extremely reliable solutions. In the automotive domain the functions of steering and braking are usually defined as safety-critical. For safety-critical functions the degree of safety is given by the reliability of the functions. Thus, the reliability of steering and braking is essential for automotive systems.

One good thing with internal hazards is that they can be found in the bounded area of the system under construction. There are a range of methods to cope with internal hazards, failure modes and effects analysis (FMEA) is commonly used and will be presented in this section. The design faults discussed here are possible to approach with improved working methods and automation of processes, i.e. fault avoidance techniques. Hardware faults must be approached by increasing reliability and introducing fault tolerance. These techniques are discussed in the previous section on reliability.

The benefits and dangers of reuse

It is common that systems are not designed from scratch but components are reused. Reuse reduces development costs as the new system can be built on the existing knowledge and investments in the old system, the development costs are spread over a higher number of produced components. The reuse of software is further seen as a good way to reduce the number of implementation bugs and improve quality, as the verification efforts of a given component to some extent may also be reused and the reused component will also be proven by hours of operational use [Shigematsu 2002].

However, the system level validation of systems with reused components is seen as a problematic issue, mainly in academia. The hazards are exemplified by the Ariane 5 accident. In June 1996 the Ariane 5 space launcher was destroyed 40 seconds after take-off. The reason for the accident was the reuse of a diagnostic software module from Ariane 4 that was not properly designed and tested for the new conditions of use. A calculation ended in an overflow because of unexpected input data and caused a system failure [Lions 1996]. In the reused component assumptions based on the properties of Ariane 4 were hidden in the design. The problem with software reuse is that it is very difficult to know if such assumptions have been made throughout the design. Fault handling for certain situations may be overlooked as the failure is assumed to be impossible based on knowledge of the original system. This may be an efficient way to go about the original design but it introduces hazards for component reuse.

In the design process of Ariane 5 assumptions were not kept clear through the use of models at a higher level of abstraction. If reuse is to succeed the changes in the system architecture need to be understood by the software engineers. Required efforts include documenting the architecture assumptions, using subcomponents that do not have system dependencies, using techniques to bridge mismatches or improving the architecture design methodology [Garland et al 1995].

Some assumptions about a truck may hold for a long time, like an assumption that the vehicle will not exceed 200 km/h at top speed. Relying on long-term assumptions improve the chances to succeed with software reuse, but making assumptions explicit by stating them in documents ensures that no accidents due to reuse will occur. Another issue identified by Leveson [2000] is the documentation of the reasoning behind specific design decisions; the documentation should include an intent specification that explicitly states assumptions made about the human interaction and operation of the system. Some designs work well together from a technical point of view but changes the behavior of the system in a way that the operator during run time do not expect, creating hazardous situations [Modugno et al 1996].

3.5.4 Hazard analysis for external hazards

Hazard analysis is an important tool for analysis of safety. When addressing external hazards one of the most difficult parts is to find the set of hazards that are significant for the system. Preliminary hazard identification should be performed early throughout the design process. For external hazards outside the system this must be performed in a brainstorming manner, for internal hazards more systematic methods like HAZOP and FMEA described below can be utilized. With identified external hazards it is possible to perform a fault-tree analysis in order to identify safety-critical components in the system [Storey 1996].

Fault tree analysis (FTA)

Fault tree analysis is a top-down method that starts with an identified set of hazards. By identifying possible causes of these hazards, and then possible causes of the identified causes it is possible to work backwards down into the details of the system. Combinations of causes are described through logical operators (such as: *and*, *or*). Using this logic the system is broken down into a tree of events for analysis. The leaves of the tree are basic events, or root causes, that are taken as inputs to the tree [Storey 1996].

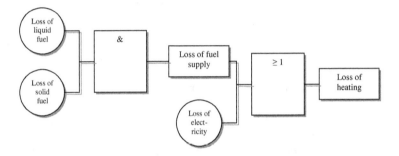

Figure 3-14 Example of fault tree (based on Storey 1996)

A simple example is given in Figure 3-14. In the example the heating is lost if electricity **or** fuel supply is lost. The *or* operation is indicated by the ≥ 1 symbol. Further, the loss of fuel supply occurs if there is a loss of liquid fuel **and** a loss of solid fuel. The *and* operation is indicated by the & symbol. Based on the defined fault tree it is possible to put up logical equations and truth tables that helps to find safety-critical parts of the system. It is also possible to add a probabilistic analysis to the fault tree in order to analyze the probability of a given hazard, and thus enable calculation of the associated risk.

Hazard and operability studies (HAZOP)

Another method for hazard analysis is *hazard and operability studies* (HAZOP). HAZOP identifies components and interconnections between them in a system. The interconnections are seen as flows and are refereed to as *entities*. Each entity has a given set of properties or *attributes*, like value or bit-rate for a computer communication system. [Storey 1996]

By studying the effects throughout the system of deviations in the attributes of a given entity it is possible to identify hazards. The analysis is performed systematically for all entities and by using a set of keywords. A sample set of keywords used in the chemical industry are shown below. For timing considerations a common addition is: *early, late, before, after* [Storey 1996].

-No	*-As well as*
-More	*-Part of*
-Less	*-Reverse*
	-Other than

HAZOP can be exemplified by considering a pipe entity between two tanks in a chemical plant. One attribute of the pipe could be flow of chemicals. Using the listed keywords it is possible to find situations to analyze in a what-if manner, for example: 'What if the flow of chemicals is missing (no)?'; 'What if the flow of chemicals is higher (more)?'; 'What if the flow of chemicals is lower (less)?' Utilizing these questions the effect of changes in the system can be systematically analyzed.

3.5.5 Failure Modes and Effects Analysis (FMEA) for internal hazards

Hazard identification for internal hazards is important in order to find safety-critical components where reliability must be strongly considered. For internal hazards the hazard analysis is performed through failure modes and effects analysis (FMEA). FMEA is a bottom-up analysis method that selects given components and investigates their possible modes of failure. For each mode of failure, possible causes are derived and also the likely consequences. The effects of the failure are determined both locally at the component and also at system level. The technique can be applied at several system levels of abstraction to achieve a refined analysis. It can be applied both early in the development lifecycle to determine safety-critical systems and components, and also later in the development to consider redundancy issues [Storey 1996]. An example of an FMEA chart is given in Figure 3-15.

FMEA for microswitch						
Ref no	*Unit*	*Failure mode*	*Possible cause*	*Local effects*	*System effects*	*Remedial action*
1	Tool guard switch	Open-circuit contacts	(a) faulty component (b) excessive current (c) extreme temperature	Failure to detect tool guard in place	Prevents use of machine – system fails safe	Select switch for high reliability and low probability of dangerous failure Rigid quality control on switch procurement
2		Short-circuit contacts	(a) faulty component (b) excessive current	System incorrectly senses guard to be closed	Allows machine to be used when guard is absent – dangerous failure	Modify software to detect switch failure and take appropriate action
3		Excessive switch-bounce	(a) ageing effects (b) prolonged high currents	Slight delay in sensing state of guard	Negligible	Ensure hardware design prevents excessive current through switch

Figure 3-15 Example of an FMEA chart (based on Storey [1996])

A problem with FMEA is that it requires implementation specific details, which means that it is *difficult to use the method early in design*. It is however an important tool when the design has progressed somewhat to preliminary implementation alternatives.

3.5.6 Modeling and formal approaches

When complexity increases, models can be useful abstractions for analysis. The abstractions in these models can be at different levels. Early in the design process it is useful to work at a high level of abstraction. Models that can be used in the first stages of the design work of a system will be referred to as *high level models* throughout this thesis. If models are utilized in a correct way there are several benefits and possibilities for verification and automation of design steps that can improve the quality of implementations [Lala & Harper 1994].

High level modeling allows functional decomposition where the actual function of each building block is defined. With a functional decomposition it is possible to arrange the system in a hierarchical layer structure that ensures that all the dependencies in the system becomes explicit. The bottom layer interfaces with the physical units while higher order layers implements abstract control and functionality. This architecture also minimizes the number of interfaces as functionality can be added in the top layer without changing things in the bottom layer. Combining layering with smart architectures gives an extremely flexible arrangement that allows quick and continuous upgrading of the vehicle system.

The beneficiary effects of modeling have been verified by a study of software quality at DaimlerChrysler. For software implementations it is important to ensure that the implemented code is equivalent to the specified functionality. This check is often performed through manual code-inspection. It has been proven that high level modeling of specifications can improve the results of code inspection. If a code inspector is forced to build a functional model of the code it is possible to find discrepancies and faults where the implementation does not comply with the specification. This method has successfully been used to improve code-inspection [Laitenberger et al 2002].

Traceability

To ensure safety properties a system must be possible to analyze. To be able to analyze a system traceability must be maintained. Traceability must be supported both within the product, through a product model, but also throughout the design process to capture the design rationale. Bracewell and Wallace [2003] propose a simple tool called DRed that captures the design rationale through the use of graphical decision trees. The importance of documenting the reasoning behind specific design decisions is also identified by Leveson [2000]. In order to avoid improper automation in a system, documentation should include an intent specification that explicitly states assumptions made about the human interaction and operation of the system.

The tool SpecTRM has been developed to support this process, organizing the intent specifications in seven predefined levels [Leveson 2002].

Well structured documentation of requirements and design decisions at several layers of abstraction are crucial to avoid external hazards due to ambiguous specifications. Proper documentation is also the key to allow reuse of designs. There exist many documentation standards. Unfortunately, many of these standards go into detail on the format of documents instead of specifying the purpose. Such standards provide headers instead of the required information content. Parnas and Madey [1995] address this issue and give an overview of required documents and also describe the required information content in detail. For reuse, it is necessary to be able to trace design decision back to requirements. The requirements should be handled through modeling and tools in order to maintain traceability. However, even though traceability is important, introducing a cost-efficiency perspective, the effort of maintaining links should be balanced by the benefits [Weber & Weisbrod 2003].

In an ideal world, a single virtual system model would be a perfect representation of a given real-world system, and thus the only necessary model for development. However, if such a model could be possible to derive, it would be overly complex and not suitable for the purpose of design. In the modeling of a complex system several models are necessary to capture all aspects. These models will to some extent be overlapping in information content. In order to ensure that no contradictions are created in the specification it is important that the different models are linked. This is, however, seldom the case in current practice. With increasing number of safety-critical functions, like drive-by-wire functionality, the need for related and unambiguous models increases [Nossal & Lang 2002].

Formal approaches

To ensure reuse it is possible to rely on formal specifications. Formal specifications are based on mathematics and cannot be interpreted in a variety of ways. Formal methods can be applied to functional specifications that are based on a theoretical model for how the system is supposed to behave. These specifications can be verified through model checking allowing deadlocks, liveness and safety issues to be considered. The functional specification forms the basis for the software implementation, which can also be modeled and checked with formal methods. The formal model checkers can also be used for generating test cases that can be applied to test the final implementation of the system [Amman, Ding & Xu 2001].

The problem with formal specification languages is that they are deemed to be too costly as the development of specifications are generally time consuming. Therefore, for cost-efficiency reasons, formal specifications must offer more than verification only and also ensure that the formally specified components are possible to reuse [Shimizu & Dill 2002].

Formal methods and modeling have been applied in the verification of a brake-by-wire system in the shape of the synchronous language ESTEREL. In this project a set of properties in the specification related to safety could be verified, also some design errors were corrected in the process [Gunzert & Nägele 2002]. ESTEREL has been used and is being used for fly-by-wire systems in the avionics industry at Airbus. In the automotive industry ESTEREL is used for verification, for example at Audi and General Motors [ESTEREL 2003].

Formal methods have also been applied in attempts to check logics and systems for potential mode confusions. The aircraft example with the 'altitude bust' described earlier has been analyzed by Leveson and Palmer [1997] and also by Rushby [2002]. By modeling both the system and also the expected mental model of the pilot it is possible to find potential mode confusions and add information to the operator that may correct the mode of the mental model [Rushby 2002].

Another method for formal specification in the context of software is proposed by Courtois and Parnas [1993]. They claim that the key to proper implementation of safety-critical system lies in the documentation and specification of the system. The proposed method is based on formal expressions of the relationships between variables. The specification can be presented in tabular form which is seen as beneficiary from a user friendliness perspective.

Code generation and Automation

One way to maintain control of dependencies and enforce strict layering policies is to define functions in high level languages like UML. With a good separation of high and low level functions it will be possible to design functions by high level modeling and then synthesize the implementation. This synthesis may include not only code generation but maybe also RTOS generation, generation of frameworks for component based development and distribution of the code on a chosen set of ECUs. Using this approach allows for efficient partitioning and system validation. It also allows for software component reuse [Schiele & Durach 2002]. The reuse of software is seen as a good way to reduce the number of implementation bugs if the system level validation can be performed in a satisfactory manner [Shigematsu 2002].

If a model based approach is supported in the design process the benefits of code generation are increased as it is expected that the models can be reused. If the model is designed as a one-off effort for a specific application there are benefits in reduction of efforts in the implementation stage, but there may also be increased costs for development and maintenance with a method that does not comply with the standard working procedures.

3.5.7 Safety and Costs

The safety attribute of a system is jeopardized by hazards with associated risks. When problems with safety are found because identified risks are too large, it is necessary to build solutions to cope with the identified hazards. Solutions for internal safety regards reliable solutions, these will be covered in the next chapter. External safety solutions require some added functionality and explicit safety functions. The explicit safety functions can be used both for active safety, trying to minimize the probability for an accident, and also passive safety targeting the consequences of an accident. Both active and passive safety measures require a development effort based on the safety analysis through methods described in this chapter, hazards must be identified and analyzed before corrective action is performed.

Some methods to improve the safety throughout design, allowing the introduction of explicit safety functions, have been mentioned. It is also necessary to reason about the costs invoked by using the different methods.

The methods regarding safety all imply a thorough analysis of the system in many ways. The hazard analysis, FMEA, functional modeling and the formal specifications all incur a higher development cost as they require extensive work and documentation. Once up and running in the system, these analysis methods have small or uncertain implications on the costs of the production, maintenance and availability of the vehicle. However, the functional modeling and the formal specifications may improve the possibility for reuse and further development of systems which improves the maintenance costs. This is further covered in the chapter on maintainability. Also, functional models and formal specifications are expected to detect and reduce the number of design errors in the system, thus improving reliability and the related availability costs. The actual extent of the impact is however uncertain.

Introducing active safety measures by adding explicit safety functions and safety supporting technology may cause higher costs, however the cost-efficiency is difficult to determine as the value of the added safety is hard to define. The implications for costs related to safety are summarized in Table 3-7.

Table 3-7 Safety methods and costs

Safety Method	Cost type			
	Development	Production	Maintenance	Availability
Hazard analysis	☹	0	0	0
FMEA	☹☹	0	0	0
Code generation	☹/☺	☺	☹/☺	0
Formal specifications	☹☹	0	☺	(☺)

3.6 Maintainability

In order to discuss maintainability it is necessary to recall that maintainability regards both repairs and also evolutionary upgrading of the system. This makes it necessary to think both ahead as well as backward in time when designing maintainable systems, forward for flexibility and backward for compatibility.

3.6.1 Compatible and flexible, cost-efficient components

The two roles of maintainability, supporting flexibility and compatibility have already been mentioned. To achieve maintainability it is necessary to enable simple and efficient reuse of designs, enabled reuse ensures compatibility and flexible upgrades. Compatibility can be achieved at component level by clearly specifying the interfaces of components. Both the structure and behavior of interfaces should be included in such specifications. New components that are introduced and are replacing old components as spare units must be compliant with other, older components in the system and this can be managed through the interface specifications. Also, future additions and changes must be planned for in the design of the current system. This does not only include upgrades and additions but also downgrades and removal of systems. It must be possible to remove malfunctioning systems that are not necessary for the vehicle to work, but still have a good basic functionality. The system should show a graceful degradation.

Systems that reuse earlier solutions are based on legacy technology from the prior system version. Several interfaces must be maintained due to backward compatibility reasons which suggest increased reuse. Also, cost-efficiency is achieved by applying the method of reuse. If designs can be reused the development costs can be severely cut and there will be more economy of scale in the development process.

Occasionally a system outgrows its legacy and must be completely redesigned. When these situations occur the backward compatibility is lost but the designers are allowed to improve the future flexibility and scalability of the system in line with the state of the art.

There are however problems with reuse as identified for software by Garlan, Allen and Ockerbloom [1995]. Within the design of a given component several assumptions are buried, Garlan et al identify four categories where mismatches can occur. Assumptions can differ for: nature of components, nature of connectors, global architectural structure and construction process. These assumptions may cause problems in the system if software is reused. An illustrative example is given by the Ariane 5 accident covered in the section on safety.

Modular solutions

Modularity has been mentioned previously in this work as a way to improve the possibility for variety management. Modularity allows the design of products that satisfies varying requirements through the combination of distinct building blocks representing stable intermediate forms.

One technique to create components or modules is to follow the principle of coupling and cohesion. In general a good module should exhibit low coupling and high cohesion. The cohesion is a measure of the internal bonds in a module, for software often referred to in seven levels. The seven levels from lowest to highest cohesion are coincidental, logical, temporal, procedural, communicational, sequential and functional. The coupling is a measure of external bonds that are often classified in: content, common (resource), control, stamp (data structure) and data coupling [Cooling 1991].

The combination of small components that are integrated to create a module depends on more than coupling and cohesion. In a platform approach many components are integrated to create a large universal module that becomes the platform. With a product line approach smaller modules are chosen. One driving factor for good modularization is the variability of the different components. If a given component need to be exchanged in order to create variety this component should not be integrated in the general platform. Further, the rate of change of components is important. If a given component will exist in a new better and cheaper version within a short period of time, the component should not be integrated in a module with components that are not expected to change rapidly. If these aspects are considered and not only functional aspects, the modularization is expected to better support maintainability and cost-efficiency [Larses & Blackenfelt 2003].

3.6.2 Layering strategies

A common way to improve the maintainability of complex systems and also to improve the possibility for reuse is to use system layering. A layer is an information hiding abstraction that specifies an interface and a given set of services that can be used by higher level implementations. However the details of the implementation of the services are hidden from the higher level accessing layers.

A common layering model for computer networks is the OSI-reference model (open systems interconnection). The OSI-model is a reference model that consists of seven layers where one or more protocols implement the functionality assigned to a given layer. The model defines *physical, data link, network, transport, session, presentation* and *application* layers. In the model layers different implementation concerns are separated, for example, the physical layer handles the transmission of raw bits over a communication link, and the data link layer collects the bits into a larger aggregate called a frame [Peterson & Davie 2000].

By utilizing layers it is possible to reduce the complexity of the complete system. A designer is only required to maintain the system within boundaries given by the layers. The more complex the system becomes the more abstract layers are necessary to cope with the complexity [Shaw 1989].

If layers are introduced the interface to the layer must be clearly specified. In software, layers are accessed through an application programmer interface

(API) that allows the programmer to use a set of services and functions through a given syntax. If inheritance and polymorphism is used it is possible to extend objects and redefine functions. This may be problematic for safety and reliability reasons by the problem of reuse, as the redesign of objects may not consider implicit assumptions in the initial design.

Behavior based functional layering

The layering of functions by using functional architecture models defines how higher level services can be organized and supplied in a practical and maintainable way. This implies a mere structuring of the implementation in a given set of abstraction levels where functionality and behavior is added along the way. Each layer hides information from other layers and act independently of these layers.

However, it is also possible to look at each layer as an improvement of a given functionality, influencing the same subsystems and changing the behavior of the system. Then the layering is not an independent structural decomposition of abstractions, but a behavioral decomposition that requires an integration strategy. A behavior-based architecture contains a set of concurrent agents competing for system resources. The global behavior of the system emerges from the interaction of the agents as they win over each other in the competition for resources. It is not structured around decomposed functions; instead it is controlled by cross-functional activities [Harper & Winfield 1994].

An example of a scheme for a behavior-based architecture is the subsumption architecture originally developed by Brooks [1986]. The subsumption architecture is a layered architecture where every layer adds some more competence to the behavior of the system. Each layer is directly connected to necessary sensors and actuators and an arbitration mechanism between the layers is implemented by a set of suppression switches. The suppression switches choose which behavior that will win, allowing priorities to be set between the layers [Harper & Winfield 1994]. An example of the architecture is given by an implementation by Connell [1989] where a mobile robot was implemented for the application to move around an office and pick up soda cans. The first two levels of implementation, the *stall level* and the *avoid level* are shown in Figure 3-16.

The *stall level* supplies motivation to move with the *trek module* that only gives the imperative "move!", the stall level also supplies the ability to change direction with the *twist* and *stuck modules*. The *avoid level* improves the behavior of the robot to change direction before collisions takes place using the *block* and *veer modules*. Higher level layers improve the behavior of the robot and implement the functionality by adding functionality layer by layer [Harper & Winfield 1994].

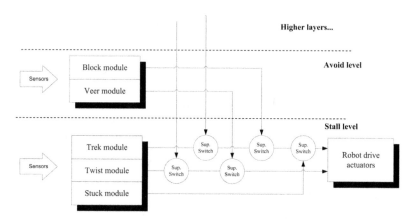

Figure 3-16 Behavior based mobile robot

With a behavior based architecture it is possible to decompose a system for design to achieve a deeper understanding of the system behavior. However, the benefit of independent layers is not achieved, and maintainability is not much improved as the integration must be redesigned according to the changed system components. Behavior based architectures should therefore not be used for system layering; however they can be used for decomposition of the services within a layer.

3.6.3 Integrating platforms and middleware

Integrating platforms and middleware architectures is an area where much research and development efforts are spent today. One purpose is to help accomplish a better reuse of developed components through separation of concerns in layers separated by the platform. It is possible to separate implementation issues below the platform level from application issues on top of the platform level. Another purpose is to improve integration of systems with specific implementations of middleware that can be applied for a given set of applications.

The integrating platforms can be seen as *stable intermediate forms,* structures that are self-supporting before they reach final configuration [Maier 2001]. Stable intermediate forms are seen as a method to improve the design of complex systems by supplying a foundation where other components can be connected and integrated.

The reuse support is obtained by the separation of parts in the system with clear boundaries and interfaces. A separation of concerns between applications and system services makes it easier to make alterations in one part while reusing the implementation in another part. Alterations can be necessary if the functionality of applications is changed. Other alterations are necessary if applications are moved to a different hardware, or a different context, that requires changes on the system service implementation level.

Middleware can be seen as a set of standardized interfaces that allows porting between functional components as well as implementation level structures, separating the concerns for implementation issues and function issues.

Middleware framework projects

Several projects have been launched with the intention to specify a standardized middleware for a specific application area, see section 2.4.4. Middleware is generally seen as difficult to introduce in automotive applications as the hardware environment is resource constrained through cost and real-time requirements. With more processing power it is expected that some standardized middleware can be used. A current effort to create an automotive middleware architecture is the Autosar project. The development partnership AUTomotive Open System ARchitecture (AUTOSAR) was formed by a broad group of players in the automotive industry in 2003. The partnership aims to establish a standard that will serve as a platform upon which future vehicle applications will be implemented. This platform includes standardization of basic system functions including bus technologies, operating systems, communication layer, HW abstraction layer, memory services, mode management, middleware/interfaces and standard library functions. The platform is intended to improve maintainability throughout the whole product life cycle, allowing software updates and upgrades over the vehicle lifetime. The platform will also provide the possibility to employ an increased use of commercial off the shelf hardware [Heinecke et al 2004].

Another effort is the EAST-EEA (Embedded Electronic Architecture) research project [EAST-EEA 2005]. This project involves European automotive manufacturers and related suppliers and tool vendors as well as academic partners. In the EAST-EEA project it is anticipated that high level programming languages will enable car designers to implement new functions or to adopt new legal requirements through the existing hardware and firmware, even in vehicles after sales and in service. Thus, the maintainability of the architecture is seen to be considerably improved. The major goal of the project is to enable a proper electronic integration through definition of an open architecture. This will allow hardware and software interoperability and re-use for most distributed hardware [EAST-EEA 2005].

Product line architectures

Platforms and middleware are seen as one way to allow parallel development of implementation and functions by using a fixed interface between function modeling and implementation [Keutzer et al 2000]. In order to implement such a platform it is necessary to find a set of general high order requirements that can be used for the specification of the services included in the platform.

Stable intermediate forms can be achieved without using a predefined middleware or complex multi-purpose platforms. A problem with rigid platforms in automotive systems is the high variability of the product. A complex platform may not be fully utilized by all products; this reduces the cost related benefits of platforms. Instead reuse based on economies of scope must be considered, allowing variants to be cost-efficiently created. For this purpose in automotive systems a product line approach is seen as a possible solution. The target of product line development is to include considerations of variability [Thiel & Hein 2002].

In a product line approach the platform is designed to be modular and scalable. To achieve this, variability must be considered early in design and cover both functional and implementation issues. It is necessary to have a clear traceability and mapping between functions and implementation to be aware of how variability requirements in functions relates to variability in the implementation. This will underline if variants will require adaptations in the architecture or if they are add on features within the product line [Thiel & Hein 2002].

3.6.4 High level modeling

The previously described methods prescribe ways to manage the complexity of the system by defining boundaries, interfaces and modular components with proper amounts of contents. Complementary to this decomposition is the support of models and modeling tools.

Decomposition with clear boundaries and interfaces is efficient for a separation of concerns in the design process. However, this method only reduces the complexity of individual design problems for subsystems within the system; in order to cope with system design and system wide complexity, abstractions and generalizations must be introduced. These abstractions are a complement to layering; the models can still be detailed covering all the aspects of the system but hiding some of the details for the system engineer.

High level modeling and automated synthesis of implementations can be used to ensure that the implementation complies with specification. The utilization of model based development improves the possibility to reuse designs and also allows verification of design changes through model based testing and simulation. The specification process becomes better documented as assumptions about the environment and the system must be explicitly modeled. If models are used the validity of documentation is ensured as the implementation is automatically synthesized from the model. Even without automated synthesis, high level modeling reduces design errors, for example in code-inspection based on modeling as shown at DaimlerChrysler [Laitenberger et al 2002].

Software has moved from one level of abstraction to the next earlier with the introduction of the, at that time labeled, 'high level programming languages' like FORTRAN and C. Development in these high level languages replaced the development in assembler code which resulted in a more readable and understandable code. The introduction of higher order abstractions also removes some of the implementation faults as the code is automatically compiled. In software, several abstraction layers have been added since, primarily in the field of IT-solutions, for example through the introduction of drag and drop Visual Basic.

A modeling language gaining status of a standard is the Unified Modeling Language (UML) patroned by the Object Management Group (OMG). As another step in the direction of high level languages and specifications OMG proposes the Model driven architecture (MDA). OMG claims that the MDA approach improves the flexibility of implementation, integration, maintenance, testing and simulation. The conclusions are based on the assumption that the models are used to generate code and implementations through automated synthesis. [OMG 2003b]

Introducing models and higher levels of abstraction may have beneficial effects in terms of reuse but also require an initial development effort, establishing the models. Further, new working methods may require investments both in tools and in training of staff increasing the size of the threshold.

3.6.5 Summary

Layering of systems supported by model based development is identified as a main technique for improving maintainability. The layering of systems can be performed in a variety of ways based on system structure or behavior. Behavior based layering should be avoided as the independence between layers is compromised; however behavior based decomposition within a layer can be useful. A layered system allows the introduction of stable intermediate forms, like middleware and platforms, which allow a separation of concerns and smaller design problems in the system.

A problem with middleware for automotive applications is the resource constrained environment that is very sensitive to introduced overhead. Middleware always include a certain amount of overhead and this may reduce its usefulness. For this reason, the stable intermediate forms should be modular in order to allow a product line variability. However, the processing capacity of embedded control units is constantly increasing and this may enable more middleware oriented solutions.

The layering of the system can be combined with the model based development. The model based development improves the maintainability in two major ways. First it ensures that documented specifications are valid against the implementation. As changes in the system must be done in the models and automatically synthesized to an implementation the documentation is the implementation. Second, changes and updates can be more easily verified in early stages of the design as the models can be tested and simulated. There is still an initial cost to establish the models before the beneficial effects can be seen.

3.6.6 Maintainability and Costs

Improving maintainability has in general little influence on production costs and availability costs; however, it reduces maintenance costs and influences development costs in both positive and negative directions.

Improving maintainability often requires an initial development effort but this effort may be rewarded with improved possibilities for reuse in addition to potential benefits in the service of the system. Implementation layering is an easy to use effort to structure the system; a more explicit layering with middleware or through the introduction of a platform may have beneficial effects in development but also introduces more overhead in the system, implicitly adding costs in production. The overhead can be avoided by instead utilizing a product line architecture and a modularization effort; modularization also improves the production through the use of a smaller subset of standardized components. However, product line architectures and modularization still requires an initial effort to create the system boundaries and interfaces

High level modeling is a powerful method that further improves the ability to reuse designs and modify a system. The creation of models however increases the overhead in development and the introduction of modeling must lead to reuse of models in order to be truly beneficial. Utilizing models in the design process are known to reduce the design errors and may have an implication on the reliability, and thus the availability costs, of the system. The implications for costs related to maintainability are summarized in Table 3-8.

Table 3-8 Maintainability methods and costs

Maintainability Method	Cost type			
	Development	**Production**	**Maintenance**	**Availability**
Modularization	☹/☺	☺	☺	0
Implementation layering	☺	0	☺	0
Middleware	☹/☺☺	☹	☺	0
Product line arch.	☹/☺☺	0	☺	0
High level modeling	☹☹/☺☺	0	☺	(☺)

3.7 Implications for automotive systems

There are a variety of methods available to improve different dependability properties of a system, each of these have a different impact on system costs. This section summarizes the conclusions in this chapter both concerning dependability and cost-efficiency.

3.7.1 Costs

Cost effects across all dependability measures can be evaluated for the four different cost types defined. Depending on the importance attributed to each cost type it is possible to draw different conclusions on what dependability methods that are most cost-efficient for a given organization.

Availability

Availability costs are obviously most improved by the methods aiming for reliability, and not so much improved by the measures for safety and maintainability. Modeling efforts for maintainability and safety may improve reliability by removing design errors, but reliability methods are dominating the availability costs. This is a well known fact, and the drawbacks of the methods for reliability in terms of other cost in production are also well known. The level of redundancy for a given system is important for availability, but is also a strong and important cost driver. Maintaining a minimum of redundancy necessary to keep availability costs at bay is a constant question for analysis and trade-off. The issue is further complicated as internal safety may require a given availability only provided through redundancy.

Maintenance

Maintenance costs are highly improved by a good decomposition of the system, both physically in the implementation and functionally. Reliable solutions may also reduce maintenance costs as less replacement is necessary, on the other hand components may be more expensive. Redundancy always increases maintenance costs as there are more components that need service.

Production

Production costs are most easy to estimate and understand. All kinds of overhead and redundancy increase production costs as more resources are built into the system. Production costs can be reduced through by economies of scale, for example by utilizing a fewer number of standard components that will be manufactured in larger series. Using fewer and modular components also improve structure related costs such as administration.

Development

Development costs are the most complex to estimate and almost all methods have implications for the development. All safety related analyses have a negative influence on development cost, but they are necessary efforts in the design of a safety critical system. Designing redundant systems for reliability also increases development costs as more efforts are necessary for the integration of the redundant systems. The structured approaches for maintainability have both benefits and drawbacks. In general they are associated with an expensive initial effort and investment that gives a return on investment in the long run. If the effort is going to be beneficial depends on the possibility to reuse the solutions for which structured working methods and solutions are applied. If the reusability is expected to be high, maintainability efforts can be very cost-efficient, if no reuse is expected they are only costly.

3.7.2 Meeting the dependability requirements

Maintainability is important for the cost efficiency. Maintainability can be designed into the system in the architecture but it can also be achieved with good book keeping and product data management aiming for configuration management. Some related methods are easy to recommend, like the introduction of system layering and modular components. These methods both improve maintainability and also have more positive effects on costs than drawbacks. A more formalized layering with tools, middleware or models may be beneficial in many cases but the methods should be introduced with care, the increased process costs should be balanced with the decreased maintenance and component costs. An important factor to consider for development costs is the possibility for reuse of the designs in the system, spreading the development cost over more produced units.

Considering that safety cannot be compromised, several of the methods supporting safety must be utilized in order to achieve an acceptable level of safety even if higher costs are incurred. Safety is ensured by thorough analysis of the system. The known methods of safety analysis require that the system is well documented and analyzable. This motivates an increased usage of computer based book-keeping and stringent product data management. Modeling may improve safety in a cost-efficient way as maintainability also benefits from modeling efforts; the positive effects are largest if much reuse is expected within the system. The cost and benefits of modeling must be evaluated from a broad perspective.

The safety analysis aims at deciding which components that must be reliable. Reliability is well known in automotive applications. Reliability in the form of availability at application level can only be increased by introducing redundancy, either in the specification of a given component or by adding multiple components. Introducing redundancy is however not cost-efficient. Redundancy drives component cost and also maintenance costs as there are more components that can break down. For automotive applications redundancy should be kept to a minimum and only be introduced when legal requirements prescribes it or when safety analysis shows that it is necessary.

Knowing the basic methods and factors that improve dependability and the incurred costs some general conclusions can be drawn. It is however also necessary to consider the market value of functions before specific conclusions can be made. The cost efficient and dependable electronic system architectures for future automotive systems will be decided by the market preferences for functionality.

Based on the analysis of dependability from an automotive perspective performed in this chapter it can be concluded that there are two main issues to deal with to achieve dependable and cost-efficient embedded automotive systems: *The control system architecture must be managed and a lifecycle process supporting this architecture must be established.* Reduced hardware costs with fewer integrated and complex nodes require that processes to analyze, document and maintain these units must be in place. *This verifies the motivation for the research performed in the CODEX project and will be further elaborated in the remaining chapters of this thesis.*

4 Managing Model Based Development

Considering that the product hardware architecture can follow different strategies for decentralization of software, such as a functional decomposition or a software centralization strategy, what does this imply? This section discusses these strategies in the light of the Eppinger/Salminen framework that suggests that the interaction patterns of the product, process and organization should be aligned. Knowledge management theories are applied to evaluate the implications by looking at complex product development as a knowledge creation process. The theories are combined with maturity models to evaluate the maturity of MBD in the automotive sector. Then, some notes to provide a lifecycle perspective are given. In the final section, a contextual perspective as well as more practical suggestions for tool implementation for MBD are provided.

Section 4.2 in this chapter is based on [Larses & Adamsson 2004]. Section 4.3 is based on [Törngren & Larses 2005] and section 4.4 of this chapter is based on [Larses & El-khoury 2004]

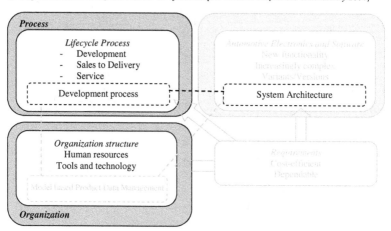

Figure 4-1 Chapter focus – Framework balance

4.1 Challenges for automotive system development

The context of automotive embedded system design provides some challenges that must be dealt with in a proper way. The systems are multi-disciplinary, growing in size and developed in parallel. This section briefly discusses these properties and what requirements they pose on the design process.

4.1.1 Multidisciplinarity

The development of complex systems such as EE architectures involves the collective effort of a variety of different specialists focusing on different aspects of the system to be built. Specialization can be defined through traditionally familiar disciplines such as mechanical, software or control engineering, or more specific roles such that of dealing with the wiring harness or safety analysis. The multidisciplinarity of the engineering effort implies that different techniques and models are needed by each of the disciplines. For example, control engineers focus on the functionality properties of the control system ignoring the implementation aspects which is the focus of the software engineers. Different models hence emerge that best suite the different focuses (see Figure 4-2). The separation of the system model into multiple views allows developers to focus on certain issues of the system at a certain development stage or for a certain purpose, while suppressing other, for the moment irrelevant, aspects.

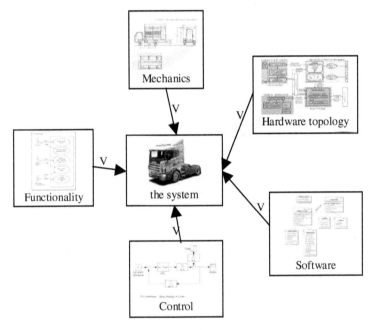

Figure 4-2 An example of some disciplines and views involved in a vehicle design

These various disciplines are integrated through the role of the system architect [Rechtin & Maier 1997]. An architect's design efforts concern trade-offs and compromise in order to reach a system-wide optimal design solution. The architect may not (and should not) have the expertise of each of the involved specialists, but enough abstract information must be collected from each of them to make informed decisions and compromises. The total cost of a system, for example, cannot be optimized by attempting to optimize the costs induced through each of the involved specialties [Rechtin & Maier 1997]. The model-based development approach needs to recognize this distribution of design effort to various specializations, disciplines or domain-specific interests, as well as the overall role of the architect.

Communication across disciplines is central in order to ensure an understanding of the common goals. Communication can be resolved by good development team integration, meetings, etc. Another, complementary and sometimes necessary technique is to utilize computer based tool support. Good model integration mechanisms allow the models themselves to be a good basis for communication.

The integration of multiple views entails that:

1. From the perspective of a given view, relevant system properties present in other views should be accessible and presentable at an appropriate level of abstraction and detail (view interfaces).

2. A change of a certain property in a given view should be reflected and propagated to dependent properties in other views (consistency).

3. From the perspective of a given view, an indication of relevant system qualities from other views is available (analysis).

Tools are needed to make this integration transparent to the designers. Currently, tool integration is often inadequate, being slow, manual and requiring redundant editors for managing cross tool information [Weber & Weisbrod 2003]. More dynamic or iterative integration is often preferable over static import/export translation of models, allowing for more flexible design processes of concurrency and incremental development. In addition, dynamic integration helps to detect and handles divergence in the understandings between the disciplines as early as possible.

4.1.2 Growing systems

One aspect of complexity is the richness of system content. This richness of content arises due to a combination of the sheer size of information available that describes the system and the relationships that exist between the information entities. This complexity is either inherent in the system or introduced through design decisions made during development.

Weber and Weisbrod [2003] highlight the complexity introduced when engineers describe specifications in an unstructured way. The manner with which the system information is modeled or presented has a crucial influence on the perceived complexity of the system. A system that is considered complex when represented in a certain fashion (for example textual models) can be quite simple when represented in a more suitable fashion (for example a physical prototype). Hence, in a model based development approach, the choice of models is important in reducing system complexity. While inherent system complexity cannot be reduced, good models help in managing this complexity and reduce the unnecessary information during design.

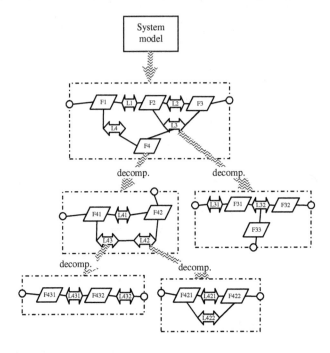

Figure 4-3 Hierarchical decomposition of a system model into parts

Mechanisms of abstraction (information hiding) and utilization of multiple views help in managing complexity. The organization of system information into separate models helps to reduce the amount of information the user needs to interact with. Furthermore, hierarchical decomposition plays an important role in providing the ability to look at the model with various levels of detail or abstraction, and hence reducing complexity [Shaw 1989]. Figure 4-3 illustrates such a hierarchical decomposition of a system model. Related information is grouped into parts that are further grouped into larger parts until the aggregated information can be easily understood. This allows the user to either focus on the internals of a part while ignoring the part's surroundings or treating a part as a black-box and concentrate on its surroundings. Design decisions can be raised from domain-specific specialists to the more generalist level of the architect as complexity increases. The decomposition of a system into parts also reflects the development process in which the effort of developing a complete system is broken down to developing its parts by different teams.

4.1.3 Concurrency

The involvement of multiple disciplines in the development of a system is most efficient when design decisions can be taken concurrently. Concurrent engineering is often referred to as parallel work in development, production and marketing but it can also concern parallel work within product development, for example in mechanical and software engineering [Prasad, Wang & Deng 1997].

The use of multiple views, hence the separation of design concerns, allows parallel development of the models. However, for this to be successful in model-based development, appropriate model integration is necessary. A designer in a certain view should be able to receive information about the system status in another view at any given point in time. This suggests that the different views should be integrated through a global, centralized product model. In addition, the mechanism to maintain consistency between the parallel development efforts should also be applied concurrently. This is an additional requirement on model integration, arguing that the consistency checks should be dynamic. It is hazardous for concurrent development, if system consistency checks are performed later in the process causing redesigns of the system to remedy inconsistencies. Concurrent engineering must also ensure access management to avoid divergent parallel design tracks of a given part [Prasad, Wang & Deng 1997].

Utilizing a centralized product data management (PDM) system may be beneficial to support concurrency. Tools/models can dynamically communicate and share information between each other through the system with stored 'meta-data' about the product and available views. It allows the sharing of product data across domains and over the product lifecycle. For example by reusing product data produced in early stages of design for manuals for production and maintenance. Further, it manages the access, consistency and integrity of data [Bryan & Sackett 1989].

Considering the system lifecycle, access to the design decisions' evolution is as important as the final decision. This history helps in tracking changes and their influence on the system performance and correctness. Mechanisms should be available for the designer to browse through the system development history, tracing solutions to requirements. The management of links providing this traceability requires resources and a selection of what links to maintain is necessary in order to be cost-efficient [Weber & Weisbrod 2003].

4.1.4 Software allocation and hardware topology strategies

In section 2.6, two main hardware topology strategies were acknowledged related to the distribution of software. With growing functional content, systems can either grow in number of components, each carrying a defined piece of software, or in internal complexity of components, with a higher number of software components. The first traditional solution allows functional decomposition and division of responsibilities, but also requires an increased effort in system integration. The number of components in the system will grow rapidly with increased functionality as recognized by the fact that the current BMW 7-series uses 70 ECUs and 2 km of wiring. The latter solution, more in line with current trends, reduces the number of ECUs in the vehicle but also makes the remaining ECUs much more complex. BMW aims in this direction and in 2003 claimed a strategy to halve the number of ECUs in 5-10 years. The integrated control units will then become a shared component for a larger number of developers, developing different functions. This strategy requires more detailed knowledge about each component and also a well defined design process and integration mechanism to allow concurrent engineering on a shared hardware platform.

4.1.5 Aligning product architecture, process and organization

The framework of Eppinger and Salminen [2001] was introduced in section 1.1.5. The model is illustrated in Figure 4-4 as a replication of Figure 1-6. According to Eppinger and Salminen [2001] the three domains, product, process, and organization must be considered when managing complex product development. It is expected that industrial firms in which the interaction patterns across the three domains are well aligned will outperform firms for which the patterns are not aligned.

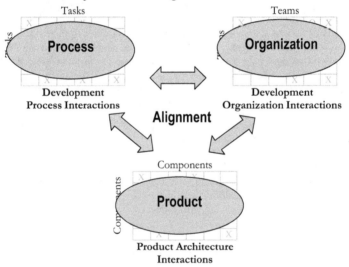

Figure 4-4 The Eppinger-Salminen framework

The interaction patters are found by looking at the relationships created by decomposition. The decomposition of the product generally refers to the physical decomposition into hardware entities. Depending on the software architecture and the allocation of software onto hardware the interactions in the product architecture will be defined. This means that the design and allocation of the software will also define the required process and organization. Assuming the possibility for two product architecture strategies for electronics as discussed in section 2.6, what do these imply for the corresponding organization and process? This question will be investigated in this chapter.

With a functional decomposition strategy for hardware and software it is relatively easy to organize one group for the hardware and one group for the software for each ECU (which carries a limited functionality). Considering a stronger integration strategy several software groups will share the hardware platform and ways to cooperate within the control unit must be found. With several dependencies between function designers, software developers and hardware developers, proper process measures must be found. The question is what these process measures look like? To provide an answer it is possible to peek at knowledge management theories.

4.2 Product development as a knowledge creation process

To align the development process with the product and organization, it must be understood how product development can be approached in a variety of ways, emphasizing either art or engineering. Theories of knowledge management provide a good framework for further understanding of the engineering process. Product development can be seen as a process of knowledge creation where the created knowledge is embodied by the final product. With this view it is possible to address the development process with theories of knowledge management. This section first introduces some theories from this field and then applies them to the design of automotive embedded systems.

4.2.1 Knowledge and the model of knowledge creation

As an introduction some concepts related to knowledge must be defined. There are several suggestions for hierarchies of information quality. Often three different levels are defined: data, information and knowledge. The differentiation is based on the "notion that structure contains more information than chaos" [Sveiby 1996]. This means that aggregation, analysis and interpretation increase the value of information.

The lowest level of information is data. Data has no relevance by itself, but merely states objective facts. Most organizations need and collect data. Data is easily collected, structured and stored but is not useful unless it is transformed into information [Davenport 1997]. Information is data that has been given some sort of meaning. Information can be described as a message between a sender and a recipient. Data becomes information only if the recipient finds the material useful [Davenport & Prusak 1998]. When information is analyzed and accepted it is converted into knowledge.

Knowledge is however more difficult to define. According to Nonaka [1994] knowledge is "justifiable true belief" and information can affect an individual's knowledge. As long as there is skepticism or the slightest doubt about the truthfulness of a piece of information, it remains to be nothing but information. When the facts or skills are genuinely adopted and believed in, the information is converted into knowledge [Nonaka & Takeushi 1995].

It is possible to distinguish between Tacit and Explicit knowledge. Tacit knowledge is "*knowledge which is non-verbalized or even non-verbalizable, intuitive, unarticulated.*" Explicit knowledge on the other hand is "*specified either verbally or in writing, computer programs, patents, drawings or the like.*" [Hedlund 1994]

By separating tacit and explicit knowledge it is possible to distinguish between four types of knowledge conversion as a basis for knowledge creation [Nonaka & Takeushi 1995], see Figure 4-5. Tacit to tacit knowledge conversion is referred to as *socialization*, explicit to explicit conversion is named *combination*. Conversion from tacit to explicit is labeled *externalization* and consequently explicit to tacit conversion is known as *internalization*. Knowledge conversion can be an internal process within an individual or take place in a transfer between several individuals.

		To:	
		Tacit	Explicit
From:	Tacit	Socialization	Externalization
	Explicit	Internalization	Combination

Figure 4-5 Modes of knowledge conversion [Nonaka & Takeushi 1995; Hedlund 1994].

Conversion from tacit to tacit knowledge, socialization, is the situation with an apprentice working side by side with his master, "learning not through language but through observation, imitation and practice". Discussions by the water cooler or informal meetings are other forms of socialization where tacit knowledge can be transferred [Nonaka & Takeushi 1995]. The process of externalization, when tacit knowledge is codified into explicit knowledge, is an essential phase for facilitating transfer of knowledge [Hedlund 1994]. Codified, explicit, knowledge is much easier to mass distribute as it can be put down on paper or into electronic documents [Hu 1995]. Combination, the conversion from explicit to explicit knowledge, is used in formal training. This is also the situation in engineering when old technologies are used for new applications. New knowledge is acquired "through sorting, adding, combining and categorizing" existing knowledge. The role of databases and information technology is central [Nonaka & Takeushi 1995]. Finally, conversion from explicit to tacit knowledge, internalization, is a concept intimately related to "learning by doing". Practicing and using explicit knowledge will create an internal know-how, a "feel" or "mental model" for the practitioner as he accumulates experience [Nonaka & Takeushi 1995].

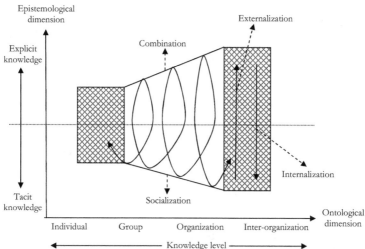

Figure 4-6 Organizational knowledge based on [Nonaka & Takeushi 1995]

Knowledge creation is achieved by cyclically alternating between the modes of knowledge conversion, referred to as the knowledge creating spiral. Further, knowledge can be amplified in the organization to become organizational knowledge. All knowledge originates from the individual but it can reach different organizational levels as shared knowledge. The knowledge creating spiral and the organizational levels are illustrated in Figure 4-6.

4.2.2 Codification vs. Personalization, choosing strategy for managing knowledge

Knowledge creation is a process that can be managed. Hansen, Nohria and Thierney [1999] recognize two distinct strategies for knowledge management in a business setting. They distinguish between *personalization* and *codification*, where personalization relies on socialization for knowledge creation and transfer, while codification uses the loop through externalization, combination and internalization. The codification strategy is based on explicit reuse of solutions and knowledge, while the socialization strategy targets tacit, unique, reinvented solutions. Codification is described as a people-to-documents strategy where the interaction patterns in the process are important, while socialization is described as a people-to-people strategy where the interaction patterns of the organization are important. The improved documentation also reduces the dependencies on individuals which may be a benefit in a rapidly moving labor market where people tend to change jobs more frequently [Hansen et al 1999].

Hansen et al indicate three factors that imply a specific strategy. The first factor is if the company pursues a *standardization* or a *customization* strategy. A standardized product increases the value of reuse and indicates that a codification strategy should be pursued, while customization is related to a personalization strategy. The second factor, *mature* or *innovative* product, underlines that a mature product based on well-known technology usually benefits from reuse and standard solutions, thus indicating that mature products should use codification. The third factor is the *use of explicit or tacit knowledge in problem solving*, the problem solving activities in the organization are expected to emphasize one of them.

Hansen et al strongly recommend a distinct choice between personalization and codification, although this choice may differ among business units within a company. Koenig [2001] has seen that a combination of methods has been successfully used at hospitals, also within business units. This suggests that it is possible to slowly evolve from one strategy to another, increasing or decreasing the codification effort as seems fit according to the situation.

4.2.3 Social and Model based development

Based on the theoretical background from knowledge management, a model of the design of embedded automotive systems can be derived. The proposed explanation model is based on a combination of the discussed theories mapped to the trinity of product, process and organization of Eppinger and Salminen [2001], illustrated in Figure 4-7. The interactions within the product architecture are considered to be given by the results of the architecture design efforts, and the complexity of these interactions is expected to increase with time. Work must be done in both the process domain and in the organization domain to fulfill Eppinger & Salminen's expectations about alignment. If Nonaka & Takeushi's theories about a cyclic knowledge creation are transferred onto this model, the focus of management efforts will alternate between the process domain (i.e. formalize tasks and methods) and the organization domain (i.e. re-organize) shown by the cyclic arrows in the figure.

Further, if the dichotomy of *personalization vs. codification* is used it is possible to make a strategic choice, placing the focus on either aspect. If integration is mainly based on social efforts it can be related to the organization, if it is mainly based on tools and models it is a process-related integration. Firms that choose a *personalization strategy* are thus expected to work more with the organization issues and less with process issues as integration mechanisms. Consequently a *codification strategy* suggests more a more model based approach, more focus on processes, and less focus on the organization. Providing terminology for product development it is possible to establish a distinction between *Social design* and *Model based development*. Knowing the difference it is possible to make a strategic choice on how to approach engineering.

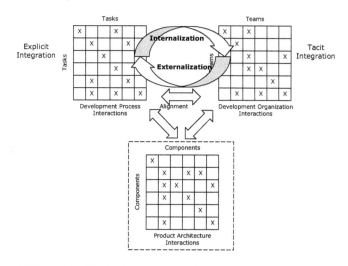

Figure 4-7 The model of Nonaka & Takeushi [1995] in the framework of Eppinger & Salminen [2001].

The dimension of personalization and codification as a knowledge management strategy can be recognized in practice in the process of aligning the product architecture, process and organization. A personalization strategy is embodied by organizational efforts such as implementing matrix and project organization structures. Process efforts, such as the introduction of development tools, formalized development models and model based development (MBD) are more in line with a codification effort. This implies that MBD is a manifestation of a codification strategy.

Social design concerns qualitative reasoning, prototyping and testing. An extreme social design process is based on trial and error where knowledge is held by individuals. Further development requires that experiences are transferred socially. The most advanced MBD approaches rely on thorough analysis by mathematically based methods and models. Such an approach often utilizes higher levels of abstraction, and provides mathematical tools for analytically comparing and evaluating alternative solutions before any implementation is performed.

Social design has benefits, being intuitive and cutting a lot of work by quickly addressing prototyping and testing. However, developing physical prototypes can be resource and time consuming, and a failing concept requires much redesign if a prototype must be abandoned. As more extensively discussed in the next section, higher complexity increases the need for a model based approach. The social approach to a complex problem is to divide it into sub-problems manageable for individuals. This means that a simple and neatly decomposed product architecture is required to ensure a simple organization structure where the sub-problems can be solved. An

MBD approach handles this complexity in the documentation provided by the models, with tools supporting analysis and integration, and maintaining the whole picture by hiding the detailed analysis in abstractions. This means that the MDB approach is less dependent on the decomposition of the product architecture.

Moreover, from the field of engineering design it is known that increasingly complex products call for extensive knowledge sharing across disciplines/functions. Such knowledge sharing needs to be provided by supporting processes and can either be implemented mainly socially through teams, or by use of MBD tools.

The Nonaka model of knowledge creation suggest that the knowledge level in the ontological domain moves up from an individual level through expanding communities of interaction that cross departmental and organizational boundaries. In an inter-organizational perspective MBD can be used as an integration mechanism. The use of CAD in product development serves as an example. At a certain point the use of CAD became interdepartmental instead of just supporting the individual engineer. Later on, standards were agreed upon that rendered an interchange of models between organizations that were using different CAD software. Now the CAD systems provide model based support for integrating all components at system level in the mechanical view of the vehicle.

4.2.4 Drivers for model based development

Nambisan and Wilemon [2000] discuss how New Product Development research has focused on integration mechanisms related to People and Process (socialization), and research in Software Development has focused on Technology and Process (codification). Naturally, the choice of integration mechanism is related to the context. MBD, however useful for integration in some mechatronic systems, is not necessarily a good integrator for the design of all mechatronic systems. In line with the thoughts of Hansen et al [1999], the *maturity* of a given product technology and a *standardization* strategy increases the possibility to reuse, and also the inclination to adopt a codification strategy embodied by MBD. The maturity is expressed for an entire industry, while standardization is expressed only for a specific market segment.

Further, complexity is often mentioned as a driver for MBD. When *complexity* is increasing, more explicit knowledge is utilized in the process of system architecting and component integration, making decisions among alternative solutions. Several inherent factors of systems make complexity management a challenging task in system development. Complexity has been discussed by many authors, for example Eppinger and Salminen [2001]. Three major factors recognized by Larses & Chen [2003] are *heterogeneity of rationale*, *richness of system content* and *conflicting requirements*. Thus, increased complexity would provide three more drivers for codification in line with Hansen et al [1999].

The *heterogeneity of rationale* behind design decisions requires more views of the system and adds to the system complexity due to the absence of a common notion of the system among the different stakeholders. The heterogeneity refers to the difference between engineering domains. A high heterogeneity is often the case in the development of mechatronic systems where domains such as mechanical engineering, electronics and software need to collaborate. Inside a system, there can be a large number of constituents with different properties and relations. The *richness of system content*, referring to the number of components and relations, increases the system complexity and increases the amount of analysis necessary for each alternative. Further, most systems need to satisfy multiple requirements such as functionality, performance, and cost. Trading, comparing and modifying requirements may be necessary for a feasible solution. For example, a performance requirement may contradict the requirements of weight and power consumption. With many *conflicting requirements* more monitored variables are needed and the trade-off activities become strenuous and more complex.

4.2.5 Empirical findings supporting the model

The model of drivers for model based development was evaluated by being applied to an empirical study on MBD of automotive embedded systems performed in the vehicle industry in Sweden. The case study was based on interviews at three different vehicle-producing companies [Adamsson 2003]. The purpose with the study was to investigate how MBD could affect integration when developing automotive embedded systems. Selected results from this study are presented in Table 4-1. The model of drivers for model based development derived in the previous section is used to explain of the results from the study.

Table 4-1 Results from Adamsson [2003] together with a classification of the companies products.

	Company ATV All Terrain Vehicles	Company T Trucks	Company A Automotive
Classification of product	Innovative product Customized product Low complexity	Innovative product Modularized product Medium complexity	Innovative product Standardized product High complexity
Results from interview study by Adamsson (2003)	MBD locally and domain-specific. Low use of CACE/CASE.	MBD locally and domain-specific. Medium use of CACE/CASE.	MBD as an integrator between different departments (domains). High use of CACE/CASE.

CACE – Computer Aided Control Engineering
CASE – Computer Aided Software Engineering

The table shows a classification of the situation for the three firms. The first row shows an evaluation of three drivers of MBD; the second row indicates the use of MBD for integration purposes. The classification of the products relates to the embedded control systems, and not the entire vehicle as a system. Entire vehicles as products are mature; however the functions

Table 4-2 Strategies and drivers of MBD.

	Drivers	Present situation
Company ATV	Customized product Low complexity	MBD used very locally Relies on personalization
Company T	Modularized product Medium complexity	MBD used locally Relies on personalization
Company A	Standardized product High complexity	MBD used semi-globally Both personalization and codification

implemented by the embedded control systems are still new and innovative. This is the same for the three companies as they operate in closely related industries.

The all terrain vehicle is the most customized product of the three, developed or adapted for each individual order. Company T utilizes modularization to create a high variety and customization with a high standardization of components, allowing each sold unit to be tailored to the customers needs through a set of standard solutions. The automotive product is even more standardized with fewer options for the customer. Thus, the automotive product is the most mass-produced of the three.

The complexity in the products is mainly evaluated in terms of richness in system content. The heterogeneity of rationale in the different organizations is similar as the electronics and software departments are studied and all three companies have mechanical products with some electronics content. All products have conflicting but different requirements, this complexity would be comparable but hard to measure explicitly. Concerning the richness in system content, the automotive company has the most electronics with the most relations, the ATV company has the least electronics content and the truck company is in between.

In company A socialization efforts were not sufficient to handle the increasing product complexity and MBD was therefore used as an integrator between organizational entities. In Company T, MBD was less used, and instead development relied on informal and tacit integration. Company ATV used MBD only for domain specific needs and not for system level multi-domain needs. The estimated level of MBD is based on the use of modeling tools. If common or linked tools and models are used by groups working in separate engineering domains the tools are regarded an integration mechanism.

The degree of tools usage for integration is an indicator of the integration strategy of the firm, as shown in Table 4-2. Considering the classification of the products, the derived integration strategies are expected. All companies operate in related industries where the maturity of EE technology is at a given level and creates a reference level of MBD. For this reason, the maturity is omitted in the table. It is possible to make a comparison with the more mature domain of mechanical engineering where MBD through CAD is used at all three companies. Companies within a given industry may diverge from the reference level determined by the strategy of codification or personalization. The complexity and standardization of the company's product are drivers to adopt a given strategy at the intra-organizational level. For embedded control systems, these drivers for MBD are strongest in the automotive company and least strong in the ATV company which is reflected in the codification strategy of the automotive firm and the socialization strategy of the ATV firm.

4.2.6 Concluding remarks

The combined theories of Nonaka & Takeushi [1995] and Eppinger & Salminen [2001] provide an interesting model of management of engineering organizations. It is possible to recognize from this discussion that a strategy for design management can be chosen, and that the focus of integration efforts will change in response to the drivers. Either a personalization strategy can be pursued performing social design, or a codification strategy with model based development. If the selected strategy matches the drivers the company is expected to perform well.

However, adopting a different strategy is not only a matter of a strategic decision on where to allocate resources. For example, introducing model based development has strong influence on the required work process. This may be problematic as established work processes - *the process heritage* - may conflict with the new process requirements. Any new model concept and related tools need to adapt to current practice and evolve gradually as they gain acceptance in the organization. Simple models and formalization of documents provides a first step towards model-based integration. The strategy, but also the process maturity, must be aligned to the drivers. This alignment is further elaborated in the next section, that also considers the maturity of tool support.

4.3 Maturity of model based development

To further develop the idea of a codified, model based development (MBD) process it must be understood what MBD is about and how it is applied in different domains?

The concepts of model, model types, model formality and model based development are defined in section 1.1.4. Model based development (MBD) is related to formal symbolic models and thereby defined as:

development based on abstract representations with predefined and documented syntax and semantics, supported by tools.

4.3.1 Evolution of model based development practice for embedded control systems

To cope with the system complexity of embedded control systems efficiently there has over the years been a constant trend to raise the abstraction levels at which the systems are being programmed and modeled, see Figure 4-8 which illustrates this for programming languages. More than 20 years ago the programming of embedded control systems (ECS) was predominantly carried out using assembly languages. Then the paradigm shifted to high-level programming languages with the idea to provide programmers with more powerful tools. These tools would relieve the burden of knowing the implementation hardware in detail, thus allowing for the possibility to work more efficiently by focusing on the applications. During this paradigm shift, concerns were raised whether the compilers would be able to produce efficient and reliable code. Entering the paradigm of model based development, the same concerns are now being raised with regards to code generation from models.

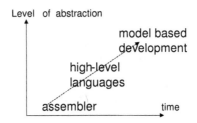

Figure 4-8 Increasing the level of abstraction

For embedded control systems development in the automotive industry, model based development is in many areas already the standard design approach; however, the adoption varies between different companies and subsystems. Computer aided engineering (CAE) tools supporting modeling, simulation and rapid control prototyping (RCP) largely facilitate development even without available mechanics and electronics hardware, and provide means for control system verification and validation in the lab. Applications are increasingly developed using model based development environments such as UML and Simulink/Stateflow. The use of code generation has increased significantly only over the last few years in the vehicular industry. For example, Volvo Car Corporation is using Simulink models in the design of power train controllers including simulation and rapid prototyping, code generated from the models is used in the final product [Lygner 2002]. Model driven testing is also becoming increasingly used, where one example is the use of hardware in the loop simulation for both subsystem as well as system integration testing. In a hardware in-the-loop simulator, the computer control system environment (i.e. the vehicle, road, driver actions as well as other relevant environment entities) are simulated in real-time, enabling system testing [Törngren & Larses 2004].

One way of reducing the failure probabilities of software in complex systems is to reduce the manual labor and individual freedom included in the development chain. This has led to several initiatives to increase formalization; for example, guidelines for C-coding have been issued by the automotive software organization, MISRA, and guidelines for modeling are being developed. MBD has an important role in this dependability effort; code generation from models is seen as one way to improve software quality [Larses & Törngren 2004]. In addition, incorporation of state of the art analysis techniques such as model checking into CAE tools is slowly emerging [Ranville 2004]. Advances in formal methods in computer science are slowly making their way into tools and industrial usage, enabling verification of software correctness prior to deployment.

It should be apparent that there are similarities but also differences in emphasis between modeling and programming languages. One common denominator for programming and modeling languages is the desire to provide readable and flexible descriptions. However, while programming languages are mainly intended to support detailed design in terms of constructing programs, modeling languages, on the other hand, provide abstractions of complex systems with the main purposes of describing such systems and/or to be able to analyze such systems. These different purposes (analysis and synthesis) are reflected in the provided abstractions/entities and inter-abstraction relations. The use of model transformations, and specifically synthesis through code generation, makes the gap between the two concepts smaller.

CAE and MBD in a mature discipline; illustrated with mechanical engineering

CAE support for mature disciplines is characterized by not only the availability of tools for handling of symbolic models, but also by the availability of supporting theory and processes for MBD, including modeling and analysis guidelines, as well as standards for model representation and exchange. These aspects are here illustrated for mechanical engineering. The term Computer Aided Design (CAD) was coined already in the early 1960's in mechanical engineering, in response to the need of dealing with geometric shapes and analysis of their behavior. The progression of CAD technology has since then developed very far and now covers advanced three-dimensional so called solid (or volume) modeling and support for finite element (FEM) analysis. The current generation of CAD-tools provides associative, parametric and feature-based modeling. Behavior models can be associated with topologies (geometry) and support for CAE assisted optimization of more complex systems is on its way [Sellgren 2003]. Solid modeling creates a kind of virtual reality for machine design, with support also for rapid creation (prototyping) of physical objects and with direct connections to computer numerically controlled (CNC) machining. Overall design methodologies as well as detailed methods and guidelines have been developed [Pahl and Beitz, 1984; Sellgren, 2003]. With the evolution of CAD and related CAE tools followed the need to be able to interchange models between different tools and different generations of tools. The ISO 10303 standard, referred to as STEP (STandard for the Exchange of Product data), has the purpose to facilitate data transfer between different information systems. STEP is mainly used in mechanical engineering by CAE and PDM tools. It includes descriptions of geometry and product structures, and is being extended to other domains. Associated modeling and product specification languages include EXPRESS, SGML and XML. Attempts to extend STEP to systems engineering data are under way [Nasa 2004]. Such efforts are of high relevance for embedded control systems.

The immaturity of using models as an integration mechanism

Several disciplines and specialists are involved in the development of mechatronic systems. While model based development is widely used within some engineering disciplines, supported by different modeling techniques and tools, integration between disciplines is still mainly relying on communication between engineers and extensive and labor intensive testing during systems integration. Examples of techniques where models today are used for integration purposes include *co-simulation* (where two tools exchange data to coordinate simultaneous simulation of different models), *code integration* (relying on code generation from different tools), *model import/export* between tools relying on exchange standards, and *super-modeling* approaches where the intention is to cover as many aspects as possible within one, more or less integrated, framework. The integration of heterogeneous models is a research area and the maturity of commercial computer based applications is low.

Introducing system-level MBD as an integration mechanism is subject to a range of requirements. The systems are multi-disciplinary and complex, the design process is concurrent, and the models must be complete for analysis as well as support efforts to search the solution space for optimization. These requirements suggest that a multi-model approach is necessary. However, using multiple views also introduces some problematic issues. The modeling process and tools must ensure *consistency* and *traceability* of information in the models. Some problems are remedied by using *centralized data storage*. Besides coping with information in the dimensions within and between domains, the third dimension of time must also be considered to maintain the development history and record changes. The choice of not using multiple views limits the size of the systems that can be considered. A simple tool without support for multiple views can give good results in an isolated case, but a more extended tool can be integrated in a seamless tool chain and reused. Complex multi-view tools have some drawbacks, but they are more easily reused [Larses & El-khoury 2004]. These aspects are further discussed in section 4.4.

4.3.2 Process maturity models

There exists a variety of process maturity models to assess and improve organizational capabilities. The purpose of these is to evaluate the appropriateness of a given work process by establishing a maturity level related to some reference model. A well known and widely spread model is the software capability maturity model (SW CMM) from Carnegie Mellon [Paulk et al 1993]. A related model is the systems engineering capability maturity model (SE CMM) [Bate et al 1995]. Some years ago the CMM evolved into the CMMI (Capability Maturity Model Integration) [CM-SEI 2004]. In both models there are five explicit maturity levels. In the SW CMM there are 19 key process areas, each associated with a maturity level. If the process areas of a given level are mastered, the organization achieves this overall maturity level and the process areas of the next level are considered. In the SE CMM there are 18 process areas in the three categories of engineering (technology), project (process) and organization. It is expected that increased engineering maturity should be supported by project and organization process maturity. The SE CMM is different from the SW CMM as all the process areas are evaluated separately; there is no general maturity level instead a maturity level is assigned to each of the areas in the SE CMM [Sage & Lynch 1998].

The five steps of maturity in the SE CMM are quite similar to the levels of the SW CMM. In the SE CMM at the initial, *informal*, level all activities are performed in an ad-hoc and undisciplined fashion. In the first step upwards, at the *planned and tracked* level, a process is established and requirements management is introduced. At the third level, *well defined*, the process is understood and standardized within the organization. At level four, *quantitatively controlled*, the process is analyzed and measured to allow active management. At the highest level, *continuously improving*, the organization is constantly improving and adapting to the changes in the environment or in the goals. The stages of the maturity process have been empirically recognized and also used for guiding quality work in the automotive industry [George & Wang 2002; Shigematsu 2002].

4.3.3 A framework for process maturity evaluation

In this section a framework to evaluate process maturity is developed. The framework uses the separation of social design and MBD established in section 4.2, combined with the basic ideas in the CMM. The CMM provides a useful, however very detailed, framework for process development. For the purposes of understanding MBD maturity a more general and aggregated evaluation of these aspects is more suitable. The five levels of maturity can be introduced for a more narrow set of capabilities concerning model based development. Relating to the Eppinger & Salminen framework we find that this process should be aligned to both the product and the organization as illustrated in Figure 4-9. Assuming a market driven process the product architecture would provide drivers for a given process maturity that must be supported by the organization resources. It is assumed that each of the three can be evaluated quantitatively by different measures, and that the maturity level of the process should match the level of the drivers to improve efficiency. The process maturity is enabled by the level of available resources.

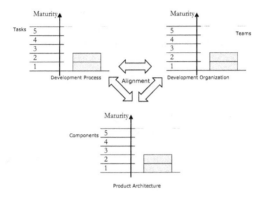

Figure 4-9 The alignment of the process context

The three aspects can be separately evaluated for social and model based development. The social and model based approaches are not mutually exclusive, but rather complementary and exist in parallel. Nambisan & Wilemon [2000] establish that product development entails people, process and technology. In our framework the social process is based on the people dimension as the main supporting organization structure. The model based process rely more on the technology dimension as the supporting part. Maintaining a high maturity for both of these processes is resource intensive and a proper balance must be established. The optimal allocation of resources among these two aspects is indicated by the product drivers introduced in the previous section.

The maturity level of model based development should relate to the complexity of the product and the possibility to reuse designs. A too high process maturity suggests that the cost-efficiency of development can be improved. A too low process maturity suggests that the organization will have problems to maintain the quality of the product. The level of support is an enabling factor that indicates if the required maturity can be easily met or if new tools or processes must be invented to cope with the process needs. The availability of supporting tools in the organization is affecting the possibility to have an efficient MBD process. For the social process the number and competence of people are sufficient provisions of organizational support. An illustration of the final framework with six aspects is shown in Figure 4-10. In the figure the five reference levels of the CMM are used, both for the social and the MBD aspects respectively, and also for the drivers and supporting technologies. The actual evaluation to quantify each of the different aspects is elaborated below.

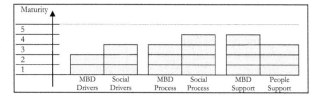

Figure 4-10 A model for contextual design process maturity

Product Drivers

Naturally, the choice of development approach is related to the context. As described in section 4.2 five drivers for MBD can be identified. Product *maturity* (not to be mixed up with the maturity in the CMM model), product *standardization* and *complexity* of the product are recognized, where complexity consist of three aspects; *heterogeneity of rationale* requires more views of the system, *conflicting requirements* call for more monitored variables and *richness of system contents* increases the amount of analysis necessary for each alternative.

In our model we suggest to quantify the MBD drivers by a combination of the proposed driver factors. The required MBD maturity level is quantified by simply adding one step for each existing driver, ideally the quantified steps would correspond immediately to the CMM levels. A *mature* product would add one step towards MBD, as would well developed standards in the area. A small product (*richness of system content*) with simple requirements (*conflicting requirements*) would rather increase the drivers for a social design approach, omitting the addition of these two steps. To evaluate the drivers in a binary manner like this is of course a simplification. For a more accurate evaluation the influence of each driver should be more thoroughly examined, for example the applicability of a given driver could be described by a fraction.

Also, the absolute need for process maturity is given by the level of competition in the considered industry and the indicated steps require a multiplication with a scale factor. In an extremely competitive industry both the social and model based drivers for a mature process may be high; the influencing factors (complexity, maturity and standardization) only show the relative distribution of process maturity requirements.

Process maturity

An evaluation of the general social and MBD process maturity levels can be derived from the specification of levels in the SE CMM. To reach each level, the proper modeling techniques and abstractions of the level must be understood by the developers. If product development is taking place, the first maturity level, *Informal* is immediately achieved. In the MBD process the use of models is non-regulated and individually used by engineers. To progress to the *Planned and Tracked* level the process must include allocation of resources and the assignment of responsibilities. The process must be documented and training in the process must be provided. The work products must be maintained under version control and compliance of the work products to the applicable requirements are performed. For MBD the second stage of maturity would require that tools are provided as a planned effort by the organization. The role of models and tools are established by plans, standards and procedures. Version control is introduced. The *Well Defined* level establishes "standard processes" across the organization. The standard process is implemented by standards and procedures, and does also incorporate verification mechanisms and completion criteria. The verification mechanisms include defect reviews on given work products. The performance is measured and the data is used for improving the process. If MBD process is utilized the use of models and tools is adopted as the "standard process". An important part of this is the availability of modeling guidelines that support developers in creating, analyzing and managing models.

At the fourth level, *Quantitatively Controlled*, detailed measures of the process performance are collected and analyzed. Quality goals are established and the performance is closely monitored. In the MBD process quantitative measures for the quality of models are established. The models must be possible to verify with mathematical techniques which implies that they are formally specified. At this maturity level, models are analyzed by tools as a part of the process. The fifth maturity level, *Continuously Improving*, incorporates processes to change the standard process. Changes are induced by causal analysis of defects found in the checking and verification process, new process activities are added, old activities may be changed or removed. For MBD, the process also provides guidelines and techniques to carry out multi-attribute optimization across models.

Support Maturity and Technology constraints

The level of support maturity can be derived from the CMM framework and organizational process areas. If a given support technology can perform the activities necessary to achieve a given process maturity, then the support technology inherits this maturity level. Focusing on the tools for MBD, the following support levels can be established. If commercial tools are available the first level of maturity is reached. For the second level of maturity there must exist support services related to these tools. The syntax of the models should be checked by tools. Also, version control support must be available and possible to implement. Uni-directional tracing from the model level to source code level by code generation belongs to maturity level two. For a well defined standardized process across the organization the tools must establish support for concurrent engineering and easy exchange of models. Further some basic verification mechanisms detecting inappropriate modeling and some semantic faults should be included in the supporting technology. Tools at this level should support several levels of testing whereby not only the models but also their successive refinement and correlation with the implementation can be tested and fed back to models of previous design steps. Examples of such support includes model level simulation, software in the loop simulation, profiling and model level debugging (where the actual software/hardware behavior is reflected back to the model level), allowing a systematic approach to testing – but still not fully analytical.

To achieve the fourth level of maturity, that of a quantitatively controlled process, the supporting tools must be able to analyze the functionality and desired non-functional aspects of the models. This requires the utilization of formal models that are non-ambiguous in their interpretation. Examples of such non-functional aspects include analysis of the system robustness, performance and logical correctness. The highest level of maturity is achieved if the tools are able to measure and check the quality of models, and support model based optimization. Further, the tools must facilitate continuous improvements. For example, templates and core models that are constantly improved are used to minimize the faults introduced in the modeling process.

4.3.4 MBD maturity in practice

In some engineering domains such as mechanical engineering, MBD is extensively used while in other domains, and for integration purposes, there is less usage of MBD. This section details the situation in a few different engineering domains related to embedded control systems within the automotive sector. The selected domains include mechanical, control, software and integration engineering. A general evaluation of the maturity is given for the process, support and drivers for MBD; the maturity of the social process is omitted. For the evaluation of the drivers, each applicable driver adds one step to the expected required maturity.

The analysis has been performed across the vehicle industry mainly in Sweden. The sources for the analysis are a collection of experiences from immediate work in the industry, co-operation projects and seminars, and an interview study [Adamsson 2003]. The results represent a typical situation of an "average" company developing cars or possibly heavy vehicles. The proposed analysis is not expected to hold in general for all companies in the automotive sector, the maturity varies across and within companies and disciplines. The examples given should be taken as representative and indicative although variations can be found in practice.

Mechanical engineering and CAD

The field of mechanical engineering is a very mature field of engineering. In the automotive sector there are many established design methodologies and theories for engineers to use, further the models and tool support are also extensive as elaborated in section 4.3.1. The drivers in the automotive sector suggest a high utilization of MBD. The mechanical parts in modern vehicles are highly mature, and to a great extent standardized across models and product programs. At the same time they are highly complex consisting of numerous components and connections like welds with non-linear behaviors, and also including conflicting requirements on weight, volume and strength. The desire is also often to optimize certain or several of these, while short development cycles are extremely important. The rationale of engineers is on the other hand rather homogeneous, yielding level 4 for the drivers according to our model.

The strong drivers and technical maturity are also shown in the application of CAD-tools in the industry. All leading firms have advanced tools that can assist the engineers and designers with most of the needed analysis. Digital mock-ups reduce the need for physical prototyping, and recent design tools even incorporate virtual reality environment where the vehicle can be explored in detail [Lohmar 2004; Monacelli, Sessa, & Milite 2004]. Model based development in the shape of CAD is an efficient and unquestioned tool in the automotive industry. A summary of the evaluation is shown in Figure 4-11.

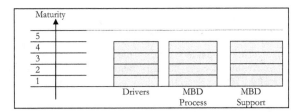

Figure 4-11 The situation for mechanical engineering in the automotive industry

Control engineering and CACE

Control theory and supporting tools have a long history of development, as elaborated in section 4.3.1. The control systems designed today are developed as separate subsystems, the trend however is towards a higher level complete vehicle control system where the different subsystems like the brakes, suspension and engine, cooperate to improve the overall dynamics of the vehicle. The high level control provides the "feel" of a car and can be used to improve performance and reduce wear and fuel consumption. This makes the control system environment complex by richness, heterogeneity and conflicting requirements. The technology is well developed and mature but there is a lack of standardization in the automotive sector, the drivers for MBD are still strong (level 4).

Matlab/Simulink is the tool predominantly used today for control engineering. A process to use the tool is implemented as a standard process across most automotive organizations (level 3). The available tools, as touched upon in section 4.3.1, provide several levels of testing and control theoretic analysis, a kind of formal verification with which system robustness can be assessed. Support for event-triggered (finite state machine) systems is becoming available (see e.g. Ranville [2004]) but this type of analysis and tools are much less utilized. In addition, although the theory is available for handling computer induced implementation problems such as quantization and time-varying delays, these theoretical results have still not propagated in use to industry. There are several short-comings for CACE tools when it comes to distributed systems implementation. Control engineering also has well developed processes, for example in the area of robotics, which implies a possibility for strong support (level 4), but in the automotive sector, the emphasis has been more on the support tool technology rather than on supporting processes. These shortcomings have been identified and several automotive vendors and research projects are working on architecture and platforms at functional as well as software levels. Therefore the MBD process for control engineering of automotive systems is evaluated to be at level 3. The situation in control engineering is shown in Figure 4-12. The mismatch of lacking process maturity is for example embodied by the questioning of the use of modeling for full vehicle simulations, not fully exploiting the potential of existing tools. It is seen as difficult to model the vehicle at proper levels of abstraction and combine different models describing different parts of the vehicle.

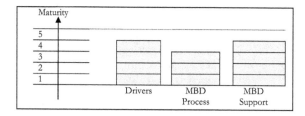

Figure 4-12 The situation for control engineering in the automotive industry

Software engineering and CASE

There are companies and suppliers in the automotive industry that are certified for level 3, but the maturity of software development in automotive industry has just about reached the lowest two levels of the SW-CMM [Huber & Näher 2004]. The use of software is not new in the automotive industry but the ability to consider networked systems with a proper process is still in its infancy. At the same time the drivers for MBD in automotive software are growing stronger. The complexity in the networks is increasing and new networking technologies are appearing. Standardizations are considered and ongoing although the standardization is still very weak. There is a lack of common platforms and a common notion of a software component. However, there is much work in the field where efforts for improved standardization include Autosar [2005] and AMI-C [2004]. These drivers suggest that a process maturity level at 3 would be appropriate [Törngren & Larses 2004].

The modeling language situation is still unstable, and there is a multitude of different formalisms and tools available [El-khoury, Chen & Törngren 2003]. The difficulty in providing tools is aggravated by the immature and non standardized automotive context. There are a variety of requirements including support for resource constrained implementations, support of variant handling and techniques to ensure high quality. Currently tools are only available to support level 2 maturity. Tools that support formal analysis are available for different but not integrated aspects and so far only used to a smaller extent in the automotive industry. The improved formalization of UML in the UML 2 specification may place tools closer to level 3, however good support for detecting inappropriate modeling and semantic faults are still lacking since UML provides very few rules and restrictions on the use of the language. To fully reach this level more experience must be collected, modeling guidelines developed and the support for analysis must be improved.

Currently, software quality problems are seen in the automotive industry in response to the lacking process maturity and tool support compared to the drivers induced by the more complex system architectures. For example, Mercedes removed 600 functions from their platform due to quality problems [Auer 2004]. The situation in software engineering is shown in Figure 4-13. The lacking process and support compared to the drivers have initiated efforts to improve the situation. Efforts in the tool area with lacking maturity in the MBD process have caused tools to be perceived as costly, inefficient and of little value. Efforts in the process area have shown the deficit in tools. The lacking MBD processes and support is also reflected by the emergence of several local tools that improve reuse of solutions locally in the organization but a cost-efficient standard process is not in place. Individual developers perceive the benefits of reuse, but no cross organizational process exists to cope with this need.

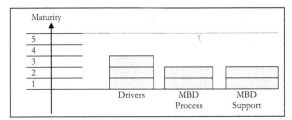

Figure 4-13 The situation for software engineering in the automotive industry

Automotive embedded systems engineering and CAI

Computer aided integration (CAI) is another field where an increased need for model based approaches are seen. The systems are becoming much more complex, specifically in terms of cross-domain interactions exemplified by automotive embedded systems. The number of global, or coordinating, functions that span subsystems are increasing. Examples of such functions include cruise controllers, traction control and vehicle stability control. This increased complexity calls for MBD although maturity and standardization are still deficient. There is a need to apply systems engineering methods of complexity management to automotive embedded systems. Several views must be provided to manage the heterogeneous context and stringent bookkeeping is required to manage the richness of system content. Methods and tools to support the documentation and analysis of function/component dependencies, qualities and other design parameters are needed. There are several types of integration required in automotive development, including architectural design, the support for co-design by interacting disciplines (e.g. control engineering and machine design; control engineering and software development) and to support integration of all the components of the vehicle. The state of practice is here varying. However, current processes for model based integration are largely weak [Adamsson 2004], yielding level 1 for the MBD process. As mentioned in section 4.3.1 there exists some tool support for model integration and also for information management but constraints and aspects affecting several domains are today mainly expressed using written documents. Tool support can be considered to be slightly above the ad hoc level. The situation is illustrated in Figure 4-14.

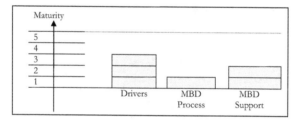

Figure 4-14 The situation for integration in the automotive industry

4.3.5 Mismatches of drivers, processes and support

As outlined here, the development process of embedded control systems is expected to increasingly use modeling and CAE tools in response to stronger MBD drivers. However, it is clear that the tools and processes are not yet as mature for ECS as for mechanical engineering. Tool and process support for model integration is also weak.

In addition, our MBD maturity model emphasizes the need to match the processes and technology support with respect to the needs – the drivers. Social methods should not be underestimated when they are applicable. The interaction between human and machine is not always efficient. Development by discussions around a large paper still has a number of strong advantages. For example, the paper format is non volatile, can be communicative, comes at a low cost and you can identify the source by hand writing style [Wolf 2002].

Balancing the use of engineering tools and engineering hours is important for cost-efficient development of commercial products. One way of evaluating the balance between social and model based development is through the suggested use of a maturity model. Measuring and evaluating the drivers and support for a given approach can be a good guideline for a successful allocation of resources. The simplified maturity model introduced here provides a good basis for reasoning about methods of system engineering. However, in order to in detail establish how a cost-efficient design process should be implemented for a given case it is necessary to do further analysis with more refined maturity models as the CMMI or the ISO15404 standard for maturity evaluation known as SPICE (Software Process Improvement and Capability dEtermination) [ISOSPICE 2005].

The topic of model based development in the field of embedded control systems is facing resistance and there exist a range of arguments. Some common arguments and pitfalls are exemplified below:

1. *Tools are costly!*

2. *Models are difficult to develop, understand, and not amenable to analysis and synthesis!*

3. *MBD is just about synthesizing code!*

4. *A model can never capture reality!*

5. *Code generation is unreliable and ECS constraints are not considered!*

6. *Overtrust in models and tools: The analysis can only be as good as the model. The garbage in/garbage out syndrome.*

7. *Too detailed models: The modeling swamp. Lacking well-defined abstraction layers.*

These objections are probably all well founded, representing experiences of misuse and malpractice of MBD methods. Several of the problems and pitfalls with MBD can be explained as symptoms of mismatches by the proposed maturity model. Issues can either be attributed to a mismatch in process maturity vs. technology maturity, or the process maturity vs. driver-induced need, as further developed below. The introduction of model based development and related tools requires assessment and understanding of the drivers, the detailed needs (what kind of support is required), and investigations of how the tools can be introduced and integrated with the process and the organization.

Process maturity vs. MBD drivers

The right level of process maturity has a strong influence on the cost-efficiency of the development process. When the process maturity is lower than the drivers dictate, development resources are used for deriving solutions that could have been reused through the application of a model based method. In addition the quality of the product may be compromised, due to insufficient analysis and testing of the new solution. Arguments such as 2-6 are typical symptoms in such a situation. If the situation is the other way around (process more mature than the drivers dictate), there is a risk that models are used for the sake of technology rather than engineering feeding argument 1 and pitfall 7.

Process maturity vs. Tool support maturity

The level of maturity in the tool support technology influences the reliability and quality of the modeling efforts. If a strong technology mismatches a weak process it is possible to achieve an over-belief in models – pitfall 6 above (such symptoms have been seen in early practices of CAE tools both in mechanical engineering and electronics design automation). Our analysis of current practices indicates potential problems here for CACE, CASE and CAI. Argument 3, not understanding the needs for co-design coming along with code generation, and 7, the risk of over-modeling, are also related to a weak process. It is very important to recognize that using mature tools in themselves solve very little. Obviously, a fool with a tool, is still but a fool. It is the analytical process maturity that ensures success in the field of MBD. A low process maturity may also create a situation of underutilization of a support tool, thus creating an image of tools as costly, inefficient and of little value – arguments no. 1 and 2. The high expectations and over-belief pave the way for disappointment with the performance of a tool. The same image may be produced by a lacking support tool maturity, in which case the conclusion is more correct – but yielding the same arguments (no. 1 and 2) creating an image of tools as costly, inefficient and of little value.

MBD drivers vs. Tool support maturity

A major problem is achieved when the need for MBD according to the drivers exceed the maturity of supporting technology. In the case where the tools are not sufficiently mature the organization must choose between performing inefficient social design, or develop in-house tools that may become obsolete within a short period of time. Thus, lack of faith in tools may originate either from lacking tools or from lacking ability to use them.

A related problem occurs if strong drivers are not fully understood. In a situation such as in the automotive with a strong mechanical engineering tradition, the decision makers may not be able to fully understand the needs for an MBD approach to handle software, control and CAI, resulting in too weak efforts on processes and supporting tools. Arguments 1 and 2 are then typically raised even if the relevant tools are available and justified from the viewpoint of the drivers.

4.4 Providing proper tool support for model based development

Model based development (MBD) has been introduced as an important advance in the design of complex distributed and embedded systems, and the maturity of an organization should match the drivers for a model based approach. However, the motivation for introducing model based methods is often without a clear consideration of needs and purpose. The positive properties of MBD under some circumstances are often used for advocating the application of the method in any context. If a model based approach is introduced with a false motivation, the delivered results will be different, and probably less useful, than promised. This section establishes a reference framework that allows contextual reasoning about modeling, and details the general framework used in the previous section. An improved understanding of the reasons for introducing model based development should make the modeling efforts more effective and efficient.

A definition of model based development is provided in section 1.1.4. A broader characterization is given by El-khoury et al. [2003], where a framework for modeling approaches is developed that is extended here. It is stated that a certain modeling approach will come along with: *tooling capabilities, modeling language(s) and its representations; modeling content; design and analysis context.* These dimensions are depicted in Figure 4-15.

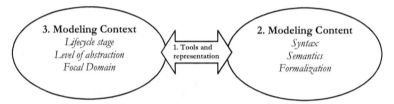

Figure 4-15 Dimensions of model based development

The context dimension considers which life-cycle stages, scope and domains an MBD approach is applied to or intended for. The context defines the appropriate level of abstraction for the modeling tasks at hand and domain boundaries of what a model describes. The modeling content considers what the models syntax and semantics actually do capture in terms of structures and behaviors. A given content should match a given context. The tooling and language capabilities provide the bridge for capturing and representing models with respect to the development context; the tools create instances of the models syntax and implement semantics and formal rules. Given the definition above, with a content, context and supporting tools, it is possible to characterize an MBD approach, to identify its suitability for a given context, and to compare different approaches. Problems with MBD can be attributed to improper modeling content for a given context, or improper tools and representations to bridge modeling content to the context.

This section focuses on requirements on the abstract representations (the models) and the tools. What do different contexts of MBD imply for models and tools? What needs must be satisfied by the modeling languages and the tool user interfaces? A framework for this analysis is proposed and discussed.

4.4.1 The Context and Purpose of MBD

Before requirements on tools for model based development can be posed it is necessary to examine the general context of modeling a bit further. Modeling and tools are utilized to support a given design activity, the purpose of the activity can be externally defined by the role of the activity in the design process. Also, the modeling activity may have a supporting role to cope with internal properties of the product that are difficult to work with without models. Thus, the requirements on the models and tools can be traced to either *external modeling purposes of the process* or *internal model drivers in the product*. Tools and models must meet these product and process requirements and supply support that is both *effective* and *efficient*. For models to be effective they should properly perform the required services. To be efficient they should reduce the resources needed to perform the activity. The reasoning is illustrated in Figure 4-16.

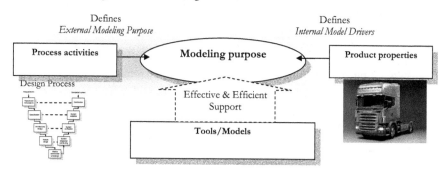

Figure 4-16 Context of modeling and tool support

The development lifecycle and external purposes of MBD

The external process purpose is based on the role in the design process of the engineer building the model. One way to address the process purpose is to consider a product development lifecycle model, such as the well known V-model [Storey 1996]. The V-model assumes that separate development process activities have a different scope and can concern either synthesis or analysis. The synthesis entails taking design decisions that add information to the product data, analysis entails verification and validation where data is analyzed to provide information and knowledge about the system under design. Also, information occasionally needs to be communicated and spread throughout the organization. In line with this, modeling either supports synthesis, analysis (verification/validation) or communication [Törngren & Larses 2004].

Through the stages of the design process the scope of modeling goes from a system wide description at a high level of abstraction to bounded parts of the system at a lower level of abstraction. For each design stage some kind of verification and testing should be performed, according to the V-model. This activity can be formalized by introducing quality gates, at each new stage a range of deliverables must be finalized and approved by a gatekeeper before the gate is passed. The number of quality gates can vary according to the needs of a given organization. The gate-keeping can also include some design decisions, thus feeding back to the synthesis. To ensure proper verification and validation the needed detail and level of refinement of the models steadily increases as each stage is finalized.

Considering the development process it is possible to establish five external modeling purposes: automated synthesis, automated analysis, gate keeping, horizontal and vertical tracing. First, a model can be the basis for reuse of solutions through *automated synthesis*, commonly exemplified by software code generation from models, where standard code patterns are reused. Also, models can be used for *automated analysis* where analysis and verification methods are supported by tools. The automated analysis can concern feasibility issues like fitting a new component in space through geometrical analysis in 3D-CAD models, or performance issues like calculating the response time for a given software task. A specific application of automated analysis would be to support the process of passing gates. For this purpose, acceptance criteria must be established and the models can ensure that information to analyze these acceptance criteria is provided. Acceptance criteria can, for example, be a minimum and/or maximum utilization of given resources such as processors and communication buses.

Further, models can be utilized for capturing product information that must be communicated across design stages as well as across organizational boundaries; *horizontally tracing* different aspects, by translating information across views, and maintaining *vertical traceability* in the design process, recording the history and progress of development in terms of variants and versions (configuration).

For models and tools to be effective they should meet the purposes proposed by the design process activities as described and summarized in Table 4-3.

Table 4-3 Activities in the design process and external modeling purposes

Design process activity	External modeling purpose
Communication – Horizontal tracing	Translation
Communication – Vertical tracing	Configuration
Synthesis – Automation of design activities	Solution reuse
Analysis – Automation of verification activities	Analysis reuse

Internal Product Drivers for MBD

Section 4.2 discusses drivers for MBD based on the properties of the product. It is concluded that an MBD approach is encouraged by a high *complexity* in the system, a *standardized* (less customized) product and a *mature* (less innovative) product technology. Complexity is recognized as the combination of the three factors *heterogeneity of rationale*, *richness of system content* and *conflicting requirements*. Based on these five drivers it is possible to derive a range of product internal purposes for MBD, each corresponding to one of the drivers, and based on inherent properties of the product that makes certain design activities difficult.

The complexity introduces three purposes, one for each complexity factor. First, models can facilitate communication for mutual understanding of a problem across different parts of the organization, *translating* product information across heterogeneous disciplines. For example, the bus-load requirements from the engineer designing the CAN-communication may need to be communicated to the engineer designing the cable harness. Secondly, richness in the system requires stringent book-keeping and documentation. The *traceability* throughout the product *configurations* must be ensured in order for the system to be analyzable. The CAN-load on a communication bus must be possible to calculate for all product variants and later changes in functional content must be traced and fed back into the analysis. The aggregated load measure must be traceable to the allocation of individual messages. Further, conflicting requirements and resource constrained systems introduce a need to make *trade-offs*. These choices can be supported by MBD with the purpose of automated analysis for system level optimization. For example, if both cabling and communication are restricted, an informed choice between another communication link and an alternative communication pattern must be facilitated.

Standardized products drive MBD with the purpose of cost-efficient *reuse of solutions*, the goal of standardization is a low cost product. Maturity supplies a driver for dependable reuse of *analysis methods*. A mature product can be reused in a dependable fashion by utilizing standard analysis methods or reusing previous analysis results, building on experience. Some testing and verification is omitted as the small changes are not expected to influence these aspects of the component. Both standardization and maturity increases the value of an investment in a model based approach as the lifespan of the validity of the model is expected to be longer. The range of drivers and derived purposes are summarized and enumerated in Table 4-4.

Table 4-4 Drivers for MBD and internal purposes

MBD Driver		Internal purpose		#
	Heterogeneity	Complexity	Translation	*1*
Complexity	Richness	Management	Configuration	*2*
	Requirements		Trade-offs	*3*
Standardization		Efficiency	Solution reuse	*4*
Maturity		Management Efforts	Analysis reuse	*5*

Linking internal and external purposes

As can be noticed looking at Table 4-3 and Table 4-4 it is possible to recognize similar purposes. The two perspectives are linked in Table 4-5. For each of the process activities that imply an external purpose for the modeling there exists an internal property that, if the property is strong, indicates a greater need to resort to modeling for this purpose. For example, modeling can be utilized with the purpose to achieve configuration. Configuration supports activities of vertical traceability in the product (i.e. understanding how the components of the solution are related). If the system to be designed is large (complexity in terms of richness in system content) there exists a stronger need for MBD, and modeling for the purpose of vertical traceability is strongly suggested. Similarly if there exists a set of standard solutions for the product, models should be used as templates, enabling automated synthesis with the purpose of reusing solutions.

Table 4-5 Relating external and internal modeling purposes

Product properties	Modeling purpose	Process Activities
Complex heterogeneity	Translation	Horizontal tracing
Complex richness	Configuration	Vertical tracing
Complex requirements	(Trade-offs) Analysis reuse	(Gatekeeping)
Standardization	Solution reuse	Automated synthesis
Maturity	Analysis reuse	Automated analysis

4.4.2 Properties of tools

In order to evaluate the appropriateness of a given tool for a given purpose, a set of attributes to describe tools must be established. Models and tools that support modeling can be classified for evaluation through a range of properties in three dimensions identified in [Larses & El-khoury 2004]. The properties concern either the formalization of *data handling* in the model, the *performance* of the tool, or the *scope of services* provided. These issues must always be dealt with in a design process but they become explicit with a model based approach.

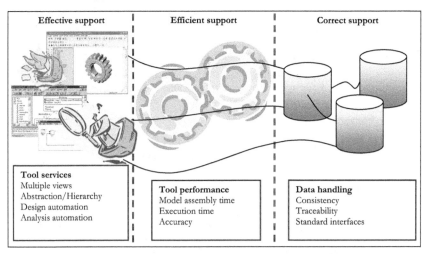

Figure 4-17 Tool evaluation attributes

The scope of the provided services of the tool corresponds to the *effectiveness*, the needed services can be provided by a single tool or an integrated toolset. The performance of the tool corresponds to the *efficiency* of the tool, while the data handling concerns infrastructure issues that must be addressed for the work with an MBD approach to be *correct*. The different properties are illustrated in Figure 4-17 and further explained below.

Tool services

The scope of the usage of the tool indicates different services that must be provided. These services can be provided by a single tool or an integrated toolset.

Multiple views in a tool (or integrated toolset) increase the ability to communicate the models across disciplines. The use of multiple views is suggested as a solution to handle multi-disciplinarity and concurrent engineering, allowing parallel design work to be performed in separate models of the system, see for example [Nossal & Lang 2002]. In addition, system complexity is reduced by allowing a domain engineer to focus on relevant properties only, without being overwhelmed with other, for the domain irrelevant aspects.

The level of details in the model indicates how complete a specification can be, how much depth and detail that can be added to a specific model. On the other hand, to cope with system wide aspects it is necessary that some *abstractions* can be introduced. A tool that supports both details and hierarchical abstractions increases the ability to navigate in the system.

Utilizing models with wide system coverage and proper abstractions, where selected *quality attributes* can be analyzed as keyfigures, the issue of trade-offs becomes much easier. Utilizing a model with a set of quality measures can focus engineering work at improvements in line with a product strategy.

Tools and models can support *automation* of engineering tasks; either by performing a chain of calculations for *analysis*, or by performing automatic application of design patterns thus automating *synthesis*. The automated steps will be performed much faster than any engineer can work, and if the proper engineering tasks are automated the efficiency of product development is highly improved. Analysis can be performed both as analytical calculations and as simulations of interesting scenarios.

Adopting Multiple Views

The interfacing of views is a mechanism of presenting relevant system properties appropriately. If performed correctly, the combination of view interfacing (horizontal filtering), together with hierarchical decomposition (vertical filtering), is a powerful complexity reduction technique. For an engineer in a given domain, the separation of concerns is performed vertically by permitting the design to focus on a single part of the system within the given domain; this vertical filtering is augmented with a horizontal filter where other domains are masked to show only the relevant information of that particular part. The modeling approach hence needs to support the designer in specifying the coupling between the relevant properties for the parts in each of the views' hierarchies appropriately. In the case where separate tools are adopted in different domains, view interfacing becomes a challenge. Figure 4-18 illustrates how associations between views may help reduce the amount of information a designer need to interact with. Figure 4-18b shows how the relevant information concerning element B1 is filtered from View1 based on the associations specified by the designer [El-khoury 2005].

Analyzing a multi-view based model can also be problematic. Model completeness is defined as the sufficiency of modeled information about the system such that a certain analysis about the system properties can be performed. Completeness for analyzability becomes harder to check with multiple views, since it becomes necessary to ensure that all the cross-domain traces are well maintained. If a given piece of information is an aggregate from other views, each of the aggregated pieces must be checked.

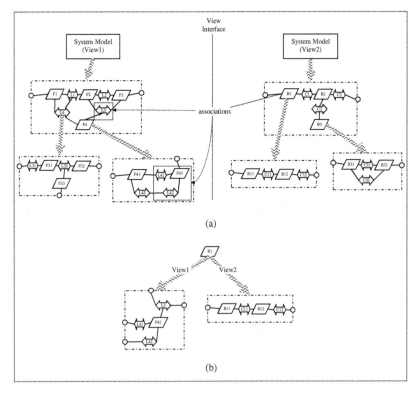

Figure 4-18 (a) Two system views together with their interface. (b) The resulting views of element B1 given the specified associations between View1 and View2.

Note that the complexities introduced by adopting multiple views are in the domain of the model and tool developers, and not in the domain of the system engineer. The simpler alternative of utilizing a single model for design is much easier from the model designer perspective. However, such an approach is expected to be overly complex for the system engineer and not suitable for the purpose of design of multi-domain systems, se for example [Nossal & Lang 2002; Larses & El-khoury 2004].

Data handling

Proper data handling is at the core of an MBD approach. As discussed in [Larses & El-khoury 2004], data handling issues refer to the possibility to maintain *consistency* and *traceability* as well as supplying a *standardized interface* to the data supplied by the tool. Consistency of models and model data is required for valid analysis.

The issue of model data consistency is complicated by the use of multiple views and concurrency in engineering. Obviously, a single engineer working in a single tool will have fewer problems with consistency than 100 engineers working simultaneously in separate tools. How the data storage is handled is closely related to consistency. Consistency, and thereby data storage has two dimensions that can be considered: [1] *space*: with data being distributed among tools or centralized and [2] *time*: using version check-in or dynamic updating.

A central source of information would better ensure the consistency across all models. This argument is supported in theory [Prasad, Wang & Deng 1997], as well as in practice [Sellgren & Hakelius 1996]. Note that using a central database does not necessarily, and should not, imply that all tools to be integrated need to use that base for the storage of their information. Such an approach is simply not practical, since not all information is necessary for other tools. Only information that is deemed relevant to other integrated models is of interest, and internal model-specific information should remain internal. A moderately coupled approach where common multidisciplinary data is stored centrally using meta-models, and complex, view specific data is stored locally is proposed by Sellgren and Hakelius [1996].

In the case where information needs to be duplicated, a mechanism of relating the duplicated/dependent information, and maintaining their dependency consistency, needs to be devised. Consistency needs to be ensured between duplicated data items (duplication consistency), as well as between data items in which one set of items is a transformation of another set (dependency consistency). It is important to notice that detailed assumptions must be made explicit as a given piece of data otherwise may have a different meaning. For example; does the number of lines of code include comments or not?

It is beneficial to handle consistency at the meta-level of the model, and hence make the mechanisms transparent to the designer; consistency should not be handled manually in the model. Automated handling of consistency and traceability requires that data can be linked through software tools. Good implementation of traceability mechanisms (including view interfaces, hierarchical decomposition and development history) facilitates the implementation of consistency mechanisms. These consistency mechanisms simply need to perform correctness checks on the established traces.

Traceability mechanisms can be based on two different approaches:

- A *meta-modeling* approach that maps all specifications back to a generally defined meta-language and can carry information across the views.

- A *translational* approach that specifies immediate mappings across each of the modeling languages and views.

Models that share common principles in their representation mechanisms through a meta-modeling approach are easier to understand. This approach is promoted by the General System Theories where a set of principles that are common between various disciplines are extracted, producing models that can be shared and understood by different disciplines through a meta-model [Skyttner 2001], this topic is further discussed in chapter 6. The alternative to meta-modeling is to explicitly specify the mapping between every pair of domains. The meta-modeling vs. translational approach are also recognized as alternatives for specifying semantics in software [Karsai et al 2003].

The consistency problem also exists in the time domain, among versions. It requires that model data is synchronized or maintained up to date. During design, distributed product data may diverge locally or in associated clusters of models. Associated data will be synchronized either dynamically, maintaining consistent data, or at given checkpoints in time when data is imported/exported. If two clusters with dependent data lack association, there may be two separate clusters with inconsistent data.

Dynamically updated data reduces the risk that different designers will have inconsistent views of the system. In their turn, model users are informed of any consistency problems as soon as they occur. Imported/Exported data may diverge between checkpoints, and requires strict validation at each check-in. Rules for this validation must be established and providing cyclic import/export facilities becomes a challenge. With dynamic integration this validation issue is avoided.

It should be remembered that proper data handling is also required even if no models and tools are utilized. In this case different specifications and documents must be properly updated and linked which is a labor intensive process.

Tool performance

Even though a tool may handle data properly, and support features that match the scope of the modeling purpose, it may still be useless if the performance does not match the needs of the modeling purpose. The tool performance can for example be measured for the *model assembly time*, the *analysis execution time* and *accuracy*.

Some tools offer a very simple interface to build models quick and easy but these models may be very time consuming to analyze, and the results may be inaccurate. Other tools produce very good results through quick analysis but require huge efforts in building and maintaining models. The profile of the tool must fit the modeling context.

A specific case of model use is real-time simulation, this case may place stringent requirement on the tool/model to be able to deliver calculated results timely. It does *not*, however, influence the analysis execution time, as the analysis execution time is equal to the duration of the simulated sequence. Other analysis methods that have a real variation in analysis execution time, for example analytical and formal approaches, may deliver the same results both faster and slower than the simulation.

Automated design steps may improve the general design process, but the performance of tools that support automation must be compared according to the parameters presented here for tool performance. A high tool performance also requires support for a concurrent process. This places strong requirements on the data handling as previously discussed.

4.4.3 Asessment of tools based on properties of MBD

This section summarizes how external and internal purposes map to requirements on tools and models. Further, there are requirements on the internal data handling that are implicit in the different requirements. Some properties of the product and design process put strict requirements on the quality of data. This section first maps internal and external purposes of modeling to tool properties and then discusses the derived needs of tools and models related to these properties.

Relating internal product properties and external process purposes to tool services

It possible to relate tool services of a model based approach both to the external purposes related to the effective role of modeling, and to the internal product properties that makes some activities more difficult. The external purposes establish the need for information handling/modeling activities in the design process. The internal purposes come from properties in the product that make specific design activities difficult, models can be introduced in order to facilitate support for these activities.

Communication mainly concerns complexity management issues. Either related to the internal purpose of *translation* in cross-domain communication (*horizontal tracing*), or related to communication along the design process at different levels of abstraction (*vertical tracing*) concerning traceability and *configuration* efforts. MBD can support translation by introducing access to a model through *multiple views* each adapted to the specific domain of interest.

A large system can be communicated by introducing *abstract constructs* that hide details of the system information. Facilitating collaboration is necessary if the designed product contains domains with heterogeneity of rationale, engineers in different domains must be able to communicate and understand each other. Kahn [1996] defines integration as the combination of collaboration and interaction. Interaction is the information exchange that is necessary for proper collaboration. To achieve this interaction the information requires translation.

Analysis and *gate-keeping* concerns activities to understand and describe the system, and is the basis for verification and validation of the system. In the analysis process the use of given models may facilitate a range of *automated* standard analysis procedures (*analysis reuse*). The gate-keeping must facilitate a *trade-off* analysis where several perspectives are considered, the weighting of the different perspectives can be supported by selecting a range of quality measures such as keyfigures linked to model based *automated analysis*.

Synthesis related to MBD concerns *reuse* activities. Reuse can be achieved by standardizing part of solutions. Through *standard solutions* synthesis can be *automated* and performed by tools through models. The synthesis supports development process management and aims at improving the efficiency and effectiveness of the design process.

Table 4-6 Relating product, process and tool support

Product properties	Modeling purpose	Process Activities	Tool services
Complex heterogeneity	Translation	Horizontal tracing	*Multiple views*
Complex richness	Configuration	Vertical tracing	*Abstraction/ Hierarchy*
Complex requirements	Trade-offs	(Gatekeeping)	*Analysis automation*
Standardization	Solution reuse	Automated synthesis	*Design automation*
Maturity	Analysis reuse	Automated analysis	*Analysis automation*

The different product properties and process modeling purposes, and their links to process activities and tool services, are summarized in Table 4-5. Each row in the table clearly separates concerns from each other; five different services, with separate drivers and separate purposes are indicated. Understanding the different purposes of modeling helps to use proper models and modeling techniques.

Mapping tool performance to design stages

One way to divide stages with different scopes is to define conceptual (architectural), embodiment (module) and detail design [Pahl & Beitz 1996]. In the conceptual stage there are many open issues and the analysis should quickly address many alternative solutions that also may differ in the way they are analyzed, the conceptual solutions may incorporate different solution technologies. However, full freedom in the conceptual design stage is only available if the design concerns a new product. Most design in practice concerns redesign, adding and modifying functionality to an existing product; in these cases the conceptual design must consider legacy issues from the design heritage of the current concept. Often the possible solution technologies are limited and the conceptual design focuses more on architectural changes in the existing concept. In embodiment design the conceptual solution and solution technology are selected and decisions on modules and system design within the concept must be established. In the detail design the solution is fine-tuned by parameters of the modules.

Different design stages are aiming at different problems and place different requirements on the performance of tools. In the conceptual stage many different alternatives must be quickly examined, models must be quick to assemble and analyze. In the embodiment phase a solution is selected and only one or a few models are needed. This reduces the requirement for assembly time but the execution time must remain low in order to prepare for the detailed design. In the detail design, parameters are fine tuned and accuracy in models becomes critical. To be efficient a tool must also provide support for concurrent engineering. This requires that the data handling is properly handled. Data must be kept consistent and easily available for all the people involved in the design process.

It is also worth noting that the modeling in different design stages may focus on different product property purposes. The detailed design may focus more on the reuse of solutions; automating synthesis by the use of standard components, thus modeling only the interconnections of standardized blocks. The conceptual stage may focus more on modeling for the purpose of translation and configuration.

4.4.4 A Detailed Framework of the MBD context

The initial framework provided in Figure 4-16 has been elaborated throughout this section. Four process activities have been identified that provide four external purposes for modeling. Five product properties that influence the modeling context by complicating design activities, and thus provide five internal drivers for modeling have been defined. Also, the need for concurrency to be efficient in the design process has been mentioned.

Tools and models are described and identified through four services. The relations of these services to the internal and external purposes for modeling are established. Three performance properties are discussed and related to the design stages in the design process. Further, some requirements on the tool infrastructure have been defined and related to different services and issues with concurrent development. The contents of this section are summarized in Figure 4-19.

It is possible to map the purpose of process activities to different services provided by tools. Further it is possible to link these pairs of activities and services to properties of the product. The properties of the product complicate the given process activities and suggest that tools and models should be used to support these activities. The proper set of services makes tools *effective*. The range of linked modeling purposes, services, activities and product properties are shown in Table 4-6.

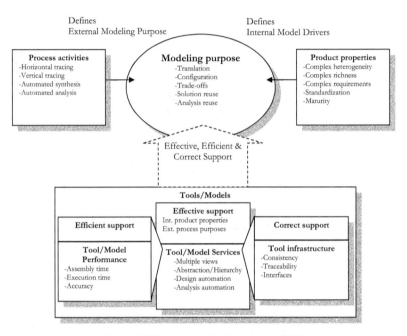

Figure 4-19 The framework for the context of MBD

Focusing further on the tools and models it is possible to recognize that the tools provide services, with a given *efficiency*. How efficiency is defined depends on the role of the tool in the external process. Also, tools are based on an infrastructure that must provide proper data handling in order to supply *correct* services. The proposed influences and required properties of tools are summarized in Figure 4-20.

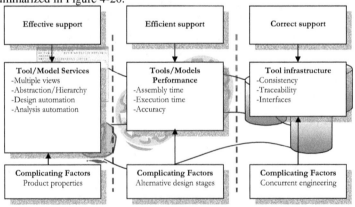

Figure 4-20 Tools and models in a context

With a focus on the purpose and context of modeling it becomes easier to choose the proper modeling and tool support, and for a tool developer it becomes easier to evaluate tool improvements if the needs of the user are well defined. A complication for introducing MBD is limitations in modeling competence and organizational maturity as well as restrictions in the maturity of tool technology; it is not certain that tools exist that can meet the requirements, if such tools do exist it is not certain that the competence to use them is available in the organization.

4.5 A product lifecycle perspective

The alignment of product, process and organization applies not only during development but throughout the entire product lifecycle. Three main processes have been mentioned, the development, the sales to delivery and the customer support processes, see Figure 4-21. It is possible to extend this classification to also include earlier phases and later phases, such as the final disposal of the system, but these phases are not in focus here. Knowing that there are different stages in the product lifecycle it is possible to acknowledge that each phase has its' own requirements on the alignment of product, process and organization. Each stage needs a set of views that filters information about the product and arrange the selected information according to different criteria. Simplifying the product architecture in one view may well complicate the picture in a different view.

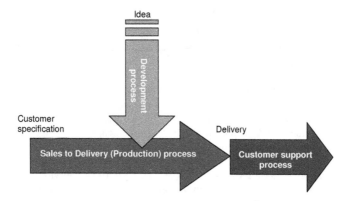

Figure 4-21 The three main processes

A codification strategy embodied by model based development can be extended using modeling and tool support throughout the entire product lifecycle. For development, models are required to cope with the entire design process from requirements to a validated and verified implementation, with an intention to automate steps in the development chain. However, this is not where the potential use of models stops. For a modular system with variants configuration management is another field where models can be useful. Configuration management can be applied in the sales-to-delivery process, where systems with alternative configurations must be assembled in a dependable manner, and also in the customer support process where each product individual must be properly analyzed according to configuration to enable diagnosis and reconfiguration. Each of the three processes usually develops its own tool support environment and builds separate models. The problematic nature of several tools and models utilized for different processes has been discussed in a context of mechanical systems. It is concluded that it is essential for efficient engineering to integrate tools, preferably through a central product data management (PDM) system [Burr et al 2004]. Failing to do so will result in losses of information between the different process stages. Traceability and easy navigation through product documentation is essential to facilitate configuration in production and efficient customer support of maintenance and repair. For a heterogeneous system, including software components, this issue is even more important.

4.5.1 Configuration management and system support for three processes

Configuration management is at the core of the integration of the three processes [Knippel & Schulz 2004]. Configuration management concerns both variants and versions, where variants are seen as the possible combinations of customer choices at a given point in time, and versions are staged changes of design entities over time.

The design process can be supported by product lifecycle management (PLM) systems with the intention to supply every step in the process with proper tool support; from requirements engineering, through development, to change management. PLM tools have emerged from product data management (PDM) tools traditionally utilized for CAD-data and mechanical engineering. UML is recognized as a supporting technology for this process in the software domain and several tool vendors supply tools that they refer to as PLM tools with UML support [Stuecka 2003]. Such PLM systems, not based on the PDM tradition, often focus on change management (i.e. versions) and have poor support for variants. In the production process, production planning tools are used that provide a mapping from a customer specification to a bill of material (BOM). The BOM provides the right versions of components needed in the production assembly of the system for a specific product variant. Although both versions and variants are covered, these systems usually have poor support for the combined configuration of software and hardware.

Management of configurations is also essential in the customer support process. It must be possible to efficiently search for the sources of faults and failures through well kept and traceable documentation. To maintain good documentation it must be possible to generate the required documents for a given variant, with given versions of components at any point in time. As contradictions can occur in the documentation for different combinations of components it is impossible to have a generic documentation set, valid for a complete product line.

Further, to provide a serviceable vehicle, challenges such as lack of training of staff in electronics and software and lack of rules and procedures for replacing and updating software require a response. If one piece of software corresponds to one piece of hardware, the product structure fits the traditional automotive structure more intuitively and the repair process becomes more tangible. If the software is distributed and hidden in the control units, new requirements are placed on work processes as well as organization infrastructure and personnel competence. To meet these requirements the diagnostic capabilities should be well developed for the product, together with a test infrastructure in repair workshops and proper training of the personnel [Näher & Radtke 2005].

The influence of product structuring

The functional decomposition and structure of the implementation are two views that are emphasized differently in different stages of the product lifecycle. In the development stage, functions are developed and fitted in the product structure. For the production stage, a simple implementation structure is the only concern, while the customer support process may require traceability in the documentation based on functions. The customer support process also desires a product structure with easily replaced and repaired components. All these aspects must be considered in the design of the product architecture.

The two mentioned strategies for product structuring, software centralization and functional decomposition, have different implications in different stages. A functional decomposition of the system with several ECUs provides a simple view on the allocation of software onto hardware. The hardware and cabling view as well as the communication view on the other hand become more complex. This makes development teams for each ECU more independent but the communication and integration group will have a more difficult job in providing cabling and communication solutions. In production the complexity is higher as more components are needed, at the same time it is easier to map a customer specification to a bill of material (BOM) that is a specification for the actual assembly of the product. For the customer support process, functional decomposition provides a more intuitive structure of components. In the extreme case one function maps to one software, that in turn is implemented in one ECU. To perform service and repairs it is easier to exchange a hardware unit, and thus the software, instead of reconfiguring the unit to adapt the software. A problem is that the combinations of hardware and software in the system must be managed at a system integration level and with a functional decomposition the number of combinations grows rapidly. Support for establishing compatibility among combinations of components must be well developed.

With a centralized product structure the software allocation and integration view become more complex while cabling may become simpler. In production the mapping of a customer specification to a product requires more careful attention to the selection of software running in the ECUs. At the same time a tighter grip on the compatibility of components is provided as an infrastructure that assures the correct software and a correct configuration must be in place.

4.5.2 Modularity in a continuous process

Managing modularity is a difficult task that is further aggravated by continuous introduction of software (and hardware) versions. In a continuous development process for a modular product, mechanisms must exist to cope with changes and to ensure that versions and variants of components are compatible, or configurable to a compatible state.

These processes must be in place throughout the product lifecycle. It must be possible to introduce updates during design time, in production and in the maintenance of a deployed vehicle. Optimally this can be achieved in a continuous manner with minor alterations in the architecture, maintaining backward compatibility. However, occasionally it is impossible to maintain compatibility and some repairs or system updates may require that mechanical parts and hardware are changed at several places in the vehicle.

Maintaining the compatibility across variants and versions is a difficult task that requires careful design and extensive verification and testing. In the SAINT project three general strategies for reconfiguration of components in a modular structure were established, hardware replacement, manual reconfiguration and automatic reconfiguration. The changed or reconfigured parts of the system must be compatible with the remaining, unchanged parts, of the system. Hardware is generally replaced to achieve compatibility. Focusing on the software a proper configuration can be established by downloading a range of versions that are verified and tested for compatibility (replacement). Alternatively different versions can be used in the same system where parameters are used to define version numbers and compatibility (manual reconfiguration). The third option is to have a dynamically configurable system that reads the versions of other components and adapts the behavior according to the actual configuration (automatic reconfiguration).

Further, the software needs to be compatible with the mechanical configuration of the vehicle and similar strategies can be utilized for this purpose. Parameterization is commonly used today but an efficient modular infrastructure for configuration management of automotive software is not yet established. Achieving a modular architecture supports a process of continuous development. However, a modular architecture is not the complete solution. An efficient and effective process with supporting methods for continuous development, production and maintenance is still a challenge.

4.5.3 Concluding remarks

This chapter initially described a set of challenges for, and characteristics of, the engineering of automotive embedded systems. In the section 4.2, two approaches to deal with the engineering challenges were provided, a model based development process was contrasted to a social development process. To guide the choice of approach drivers for MBD were established, based on studies in the Swedish automotive industry. These drivers were then used in a discussion on the maturity of model based development throughout the automotive sector. It is proposed that the extent of MBD usage must be in line with the drivers and supported by tools that are mature enough to provide proper support.

If a model based approach is to be applied, it must also be established in what form models and tools are to be utilized. Section 4.4 provided some details on the selection of models and tools by relating tool services to the purpose of MBD. It is recognized that model based development can be advocated for a range of reasons, but occasionally the wrong reasons are used, resulting in an inadequate model based development process.

The selection of proper models and tools to support the design process of a product is improved by knowing the context in which the tool and models are to be used. Organizations that try to implement a model based development approach must be aware of both their product and process needs. The proposed framework, with requirements on effective, efficient and correct services, may increase the possibility to find the proper support for a cost-efficient model based development process. To really succed, the requirements from the entire product lifecycle, with special attention on configuration and translation, must be taken into consideration.

5 An Architecture Engineering Process

Architecture design is often referred to as an art, performed in the conceptual stages of a design process. However, it can be supported by quantitative methods. In this chapter a method for architecture analysis and a method for architecture synthesis are proposed. The analysis and synthesis methods are combined to produce a complete architecture engineering process, supported by quantitative methods.

This chapter is based on [Larses 2005a; 2005b] and [Larses & Blackenfelt 2003], section 5.4 is based on [Larses & El-khoury 2004].

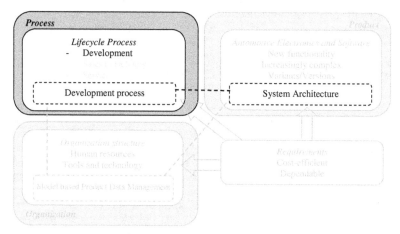

Figure 5-1 Chapter focus – Architecture design

5.1 Architecting with quantitative methods

As previously discussed, architecture design is becoming increasingly important for automotive embedded control systems. The design of the on-board computer network and choice of related components has an increasing leverage on costs and a good architecture will improve the capability to build dependable and cost-efficient systems. Assuming that the most rigid part of the system is the hardware, the choice of hardware components and technology becomes very important in order to facilitate the desired support for software and functionality, providing enough I/O-ports for sensors and actuators and enough computing power.

As discussed in section 1.1.2 the utilization of information technology can assist architecting to make a transition from art to engineering. By managing large amounts of information about the product in a structured way, more detailed analysis and evaluation of alternative architectures is possible. A problematic factor is that architectures often are designed for a future product and many of the details of the future system are unknown. Therefore models that provide some basic information about expected properties of the system are required early in the design process. In this chapter some modeling abstractions useful for this purpose are provided, further discussed in chapter 6. The models have been tried out in case studies.

Architecting concerns the process of finding bounded modules in the system. Architectures can be analyzed and compared, but before analysis methods can be employed it is necessary to develop some possible alternative solutions. To define architecture alternatives a synthesis method is required. *Cluster analysis* has been utilized for defining modules in general engineering design problems, in the case studies presented here these methods have been applied to the embedded system architecture problem. Cluster analysis in general deals with the problem to group objects with similar properties into categories. The degree of association should be maximal for objects within a cluster and minimal to objects outside the cluster. The cluster analysis provides a structure among objects in a dataset. The use of cluster analysis for modular architecture design has been explored in case studies presented in this chapter.

The design process must incorporate a trade-off activity where several requirements and aspects are considered and balanced. In an industrial context, such as the automotive industry, there are not only technical aspects of function and implementation but also strategic aspects of organization and procurement. This trading activity can be performed tacitly by engineers but also be supported by tools. This chapter shows a method utilizing mathematical tools, developed in the work with the case studies. The numbers used by the mathematical methods provide a tangible basis for discussion and design decisions. If requirements and assumptions thereby become explicit, as the alternative solutions can be mathematically characterized, the architecture can be more accurately designed to fit the intention of the system. The rationale of the design is often poorly documented, especially at the architectural level. Using a combination of keyfigure analysis and clustering based on module drivers is one way to improve the documentation of architectural assumptions and requirements. Here, keyfigures are proposed as a method for analysis and clustering with design structure matrices (DSM) is proposed for synthesis. The proposed method has been applied and evaluated in architectural work at Scania CV.

5.1.1 Concerns for architecture design

In this context, architecture refers to the structuring and decomposition of the implementation of a product in hardware and software with predefined functional contents. The architecture can be evaluated either for a specific product variant or across a set of variants and versions. The quality aspect for a given variant can be measured and captured in proper keyfigures of choice.

With a single product in mind the architecture can be optimized to this product, with variations the optimization must be balanced across the set of product variants. The modularity and scalability aspect of a product range or family of variants (*variation in scope*), together with the possibility to maintain, upgrade and downgrade the vehicle (*versions, variations in time*), must also be evaluated. For the automotive industry, products both exist in product ranges and they also change over time.

Ulrich [1995] distinguishes between *local* and *global* performance characteristics. The local performance characteristics can be optimized for the system in each module, while global characteristics can only be optimized at the system level inflicting dependencies across modules and reducing modularity. For automotive systems a balance between concerns for modular solutions and global optimization is necessary. In order to achieve this balance, a balanced scorecard approach in combination with modular clustering techniques is proposed in this thesis.

The proposed approach assumes that the functional content of a set of product variants is given. The architecture of the system is implemented to carry this functionality for all variants, and easily provide the specific content of a single product variant. The design process is expected to be *customer driven*. It is also possible to reason that functions can be designed based on a given control system hardware architecture, a *technology driven* design process. However, the automotive industry is commonly considered to be customer driven, why the assumption of such a design process is appropriate.

5.2 Modular architectures

Modular products have received a lot of attention during the last decades, because modularity is credited with a range of potential benefits for the companies that employ its principals. It is claimed that a modular product provides better handling of product variety, improved organization of development and production as well as improved handling of various after sales issues, in comparison to a product that exhibits a lower degree of modularity. Establishing a modular platform would be one way for manufacturers in the automotive industry to efficiently deal with variability.

5.2.1 What is modularity in an automotive application?

In the literature various theories to describe modular products have been suggested in order to confront the question: What is a modular product? Modularity has been defined and methods for modularization are developed based on two main viewpoints; either the product with its elements and relations (technical factors) or the purpose of the modularization (strategic factors) has been in focus. The latter, strategic perspective that focuses on the purpose of modularization describes a modular product as a product with exchangeable elements for creation of variety [Pahl & Beitz 1996]. With a technical perspective, modularity is described in terms of the product elements; it is postulated that there should be a simple mapping between functions and parts, and there should be strong links between parts within the modules and weaker links between parts of different modules [Ulrich 1995]. The ideas advocated here builds on thoughts of Blackenfelt [2001], who argues that both the purpose of the modularity and properties of the products elements and relations, referred to as the strategic and the technical viewpoint, are needed to define the architecture.

In the area of engineering design many previous studies have focused on the modularization of the mechanical domain. In this domain it is difficult to redefine modules based because of physical limitations. Functions have a straightforward mapping to the mechanical components. In this thesis mechatronic systems are studied, exemplified by vehicles, and more specifically in the context of the hardware and software architecture of modern trucks. In mechatronic systems complexity will be transferred from the mechanical domain to software and electronics that can more easily be moved and reassigned to improve modularity. Due to this flexibility it is argued that the methods used for modularization are especially suited for this type of products. Moreover, it should be noted that in the automotive industry today function growth already takes place in software and electronics rather than in the mechanics. Therefore, it is highly relevant to study how to modularize the domain of mechatronics, in order to handle the increasing complexity.

In a modular automotive product, it is desired to add, remove or change functionality by simple adding, removing or exchanging of elements of the product. This modularity must be achieved with strict requirements on a cost-efficient and dependable solution. To design a modular architecture more focus on system-level design is required to carefully define component interfaces, standards and protocols. The automotive OEM buys sub-assemblies from a range of suppliers and acts as a systems integrator. However, a modular system requires a system architect rather than a system integrator, assuming a more proactive role in the design of components [Ulrich 1995]. The OEM must take responsibility for the detailed specification of interfaces to achieve a modular system in the role of system architect with several suppliers connecting to the system. Automotive manufacturers must also consider strategic issues related to procurement, the modularization effort should include organizational and process issues that influence the ability to perform changes in the system architecture [Larses & Blackenfelt 2003]. Thus, both technical and strategic issues are of interest. The different factors that influence modularity, including non-technical aspects, are labeled module drivers and are further discussed in section 5.3.

A wide range of methods and procedures for product modularization have been suggested to answer the question: How to create a modular product? It is obviously impossible to modularize a product without knowing what modularity is, and thus methods are often derived from a definition aiming either at the purpose of the modularization or the encapsulation of the product elements. The Design Structure Matrix (DSM) has frequently been employed to describe the relations between elements of the product in order to define suitable modules [Pimmler & Eppinger 1994]. On the other hand, the Module Indication Matrix (MIM) with the module drivers of Erixon [1998] has proved interesting for describing the purpose of the modularization. These approaches have been combined by Blackenfelt [2001] and are further developed and exemplified here.

Furthermore, in the previous studies of mechanical products, using for example DSM, the module boundaries have merely been defined for an already designed product, i.e. only smaller lay-out changes have been possible when the needed parts are known. When studying products which mainly are realized by software, there is more freedom to structure the allocation of functionality late in the process, at the point when the necessary parts are known. Thus, DSM should be more suited to be used for this type of product.

In reality, it is rare that an architecture is created without constraints, from legacy and previous decisions. The existing system can be taken into account by documenting the existing relationships and required components. All the case studies performed have been forced to deal with this issue.

5.2.2 Modularity in standard automotive hardware topology solutions

Assuming that modularization is analyzed through cluster analysis, what does a given suggested clustering imply? To answer this, common ways to build hardware topologies to facilitate the desired support for software and functionality must be known, this topic is covered in more detail in chapter 2. Automotive functions implemented by electronics have traditionally been independent and therefore the control system has been built from independent components. Today, dependencies are becoming more and more complex as different distributed applications like adaptive cruise control and stability programs introduce functional relations across several control units; also simple functions become distributed as sensors and operator interfaces are shared across the system. Therefore, a modern automotive EE system is generally built as a network of electronic control units (ECUs) connected through a serial bus, where the dominating bus protocol is the CAN-bus. Other protocols are also used, introducing diversity, and the future will see evolving protocol standards. The use of serial bus communication is however not expected to change.

Modularity can be created by implementing functionally independent hardware modules that can easily be added to or removed from the control system. This is the traditional solution that has shown to be very hardware intensive, and therefore expensive. Today functions generally share the hardware space on ECUs and the trend is to increase the hardware sharing in order to reduce component costs. The intention is to simplify the hardware structure by providing capabilities to handle configuration of functions in the software. The solution to the modularization problem of electronics is however not yet final, but either way the design of the architecture is important. With independent modules the topology of the network is essential, with integrated units the allocation of software onto hardware units become another important design issue. With the sharing of hardware a new dimension of modularization needs to be considered, managing the configuration of software modules.

The system architecture should be based on the requirements from a function structure. The function structure in the case studies is defined by the communication and synchronization needs across functions. For applications concerning the driving of the vehicle this is equivalent with the control design. The control design implies requirements on which sensors and actuators are needed and also provides a functional decomposition that can be used for the structuring of the control software. With a defined control and software architecture the software needs to be deployed on the hardware. In this process two software blocks can be separated or placed in the vicinity of each other. It is possible to distinguish between three levels of software proximity. Two pieces of software can be on the same network separated by a gateway, on the same bus or in the same ECU. Software can be separated because of integrity issues; separating software reduces the risk for data contamination. Arguments for increasing the proximity are performance arguments related to communication needs. A high proximity reduces communication delays and also supplies higher bandwidth, there is a significant difference between internal ECU communication and communication between different buses over a gateway. If two software blocks are closely related in the software architecture there may be timing requirements on the implementation. In a real-time control system important requirements include worst case response time of a triggered task, communication delays as well as the period time of sampling and execution.

The trends in topology and control distribution in automotive networks reflect the underlying functional structure, where increased complexity is handled by building hierarchies and implementing sub-networks. Reilhac & Bavoux [2002] distinguish three levels of system design: the local ECU with mechatronic sub-modules, the cluster network, and the backbone and gateways. It is assumed that future network topologies will be built with such hierarchical sub-networks, and with one or a few central nodes. Based on this assumption, the results of a cluster analysis can be interpreted as a grouping within a control unit (merging two units) or as a grouping of units on a sub-bus. Weakly dependent units can be separated by gateways in the network topology.

The Scania architecture

The Scania system architecture used in the case studies provided in this chapter is built using the controller area network (CAN) protocol, and it is based on three buses separated by a control unit that acts as a gateway. The J-

1939 standard prescribes a set of messages that are used for the communication in the network. The gateway unit is called the coordinator and features some software functionality apart from the role of a gateway. ECUs with different levels of system criticality are separated by being placed on different buses which are identified by color labels, see Figure 5-2. The red bus has ECUs with the highest criticality. ECUs on the yellow bus are estimated to have intermediate criticality and the green bus have the lowest level of criticality.

A group of related functions are labeled as a logical ECU. Logical ECUs are then allocated on the physical ECUs. Not all control units are included for a given vehicle as the functionality varies with each truck. The architectural concept of Scania has separated much of the functionality onto separate physical ECUs but there are exceptions, for example the coordinator collects several functions. Moreover it should be noted that the main software of a given function may be located in one ECU, while the other ECUs are necessary parts of the function. Typically, the main function may be located in one ECU while the other ECUs are needed as interfaces to actuators and sensors.

The ECUs are generally placed on the bus with appropriate criticality given by their functionality, but there are exceptions. An ECU might be placed on a bus with higher integrity level in order to reduce excessive communication or construction penalties associated with placing it on the bus with appropriate level of integrity. Aspects for the system design include: available technology, reliability, safety, cost, sub-contractors etc. These trade-offs are currently performed through qualitative investigation efforts. The choice to place multiple logical ECUs in one physical ECU is due to considerations in line with the trade-offs.

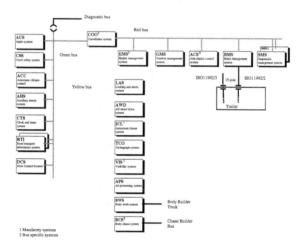

Figure 5-2 Scania ECU topology

5.3 Factors influencing modularity

The proposed method of modularization introduced in this chapter is based on a DSM (Design Structure Matrix) approach where several perspectives are evaluated and weighted together. This section expands and discusses some relevant perspectives and module driving factors. Modularity mainly refers to the physical domain, including hardware topology combined with the allocation of software. However, by ignoring some aspects the theory can also be applied on a pure software or hardware system. Technical and strategic factors influencing the modularity of the control system architecture are covered and the technical factors are specifically sub-divided into functional and implementation factors. The factors are based on the requirements from the automotive domain to build dependable and cost-efficient systems and have been developed in the line of work with the different case studies.

5.3.1 Technical dimensions

A mechanical system is mainly defined by the geometry of components. A mechatronic system containing an embedded control system adds several degrees of freedom. The same mechanical design can be utilized for different purposes and implement different functions defined by the software contained in the microprocessors of the control system. At least four degrees of freedom can be considered [Coelingh, Chaumette & Andersson 2002]: functions, hardware topology, electric signals and software and communication. Integrating the design of function, hardware and software has been discussed by El-khoury and Törngren [2001]. Each dimension inflicts relationships between parts of the product. The basis for the technical relations is the functions, but known facts about the implementation should be used when possible to provide a better analysis for modularization. Such knowledge could come from legacy systems that will not be changed, industry standards and prescribed ways of procurement indicating specific suppliers with specific solutions.

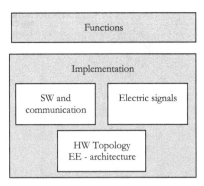

Figure 5-3 Technical dimensions of embedded systems

Function

A function can be defined as a transformation process that accepts a range of input and produces some output [Schindel 2005]. A function can react to a given event by executing a given behavior, but also work in a continuous process of input and output. Functions can be decomposed and are related to other functions.

Functions are pure abstract constructs and require some means for implementation. In the implementation, given functional blocks are assigned to components thus creating implicit functional relationships between components.

Software and communication

The software and communication dimension is the actual implementation of the functions on the platform. The software design in combination with the allocation of some hardware carrying analog information, such as sensors and actuators, is the basis of this dimension. This includes an allocation of the software onto control units that creates a specific communication pattern between the units in the network. The allocation of software is implicitly an allocation of functionality.

The mapping is very important for proper modularization of functions and hardware. A debated issue is the proper allocation of software, and the degree of centralization or decentralization of functionality. Performing thorough modularization analysis is expected to support proper decentralization decisions.

Hardware topology

The hardware dimension is defined by the positioning of hardware units and the interconnections through cabling. In the hardware view the actual signal, software and functional content of the components are ignored and design issues concern cable routing and physical space and environment of electronic units. The hardware topology is the most rigid of the dimensions and most difficult to change, therefore careful attention must be given the design of this architecture; this section proposes a range of factors to consider for this purpose.

In the hardware topology the allocation of IO-ports for sensors and actuators is included. This allocation is a very important part of the implementation of functions and the allocation of IO-ports strongly influences the functional relationships across control units in the network.

Electrical signals

In the signal dimension issues like EMC are considered together with signal to noise ratios and signal quality. For automotive networks, termination of communication links is an important issue to consider related to the signal dimension. Electrical problems can be solved by choosing different communication speeds and alternative cable paths.

5.3.2 Design approaches

To achieve a well performing system the four degrees of freedom: function, hardware, software and signals, must be aligned. Each of these domains can be altered to make the system feasible. To achieve a given functionality different implementations featuring alternative software, hardware and signal configurations can be effective.

In a design process some parts of the system are designed and some are assumed to exist. It is often assumed that the functional content is required and that the hardware topology is designed, this is however not an obvious precondition. Usually a system architecture already exists that may impose restrictions on the freedom of design, and also the functional content may not be fully known. In these situations the basic ideas of module driving factors are still valid. The results of the analysis then propose an allocation of new functionality in an existing architecture, as exemplified by the ACC case further developed in this chapter.

A system can always be implemented according to specification if the functional content in the specification can be changed or reduced. Further, a problem with communication bandwidth can be solved by changing the signal transmission rate, adding an extra cable for communication, or redistributing software (and possibly IO-ports). The choice of approach is related to the design process. For a modular solution, general problem solving should be shifted to the domain where it is most easily handled in the current process. The ease of problem solving in each dimension depends on the design process as well as the technical architecture and the organization.

Each dimension may introduce implicit dependencies in the other dimensions. Sensors and actuators can be shared or dedicated to a given function. In a software based system the function is mapped to software components and related sensors and actuators. The function defines an implicit link from input (sensors) to output (actuators) that must be implemented. If the implementation of two functions shares a sensor the two functions become related. The system modularity can be compromised by using shared resources. For example, in Figure 5-4 function A and function C are both depending on sensor X, so a change in function A may influence function C, and a change in sensor X influences both functions. Such dependencies will exist, and should be managed so that changes to improve one function will not impair the other function, functions that share resources should evolve together. Understanding the functional domain is key to the implementation of a proper mechatronic system architecture.

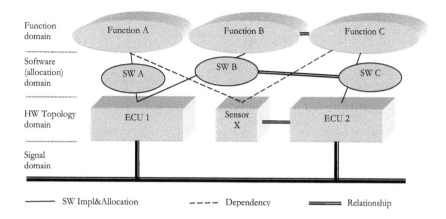

| Function domain | Software (allocation) domain | HW Topology domain | Signal domain |

—— SW Impl&Allocation   - - - - Dependency   ══════ Relationship

Figure 5-4 Relations in different dimensions

5.3.3 Module Driving Factors

Assuming that the organization, design process and function specification are given, there are a range of factors influencing the topology of the hardware architecture. Some concern strategic factors that improve the procurement and administration of the system; some are related to the functions supported by the system, the functions make different parts of the system strongly related. Other factors concern the technical implementation of the system and the feasibility of actually building the proposed system. All these factors are here labeled module drivers and can be used in a cluster analysis as further developed in section 5.5 to suggest a clustering in the hardware topology, an illustration of the drivers is provided in Figure 5-5. As mentioned previously other authors have suggested a wide range of general reasons to group elements, the factors proposed here are derived from the requirements posed by embedded systems in the automotive domain.

A given module driver can provide a relationship between two components either by classifying them according to a classification scheme, such as if the component is expected to be developed and manufactured in house or bought from an external supplier (the make/buy distinction), or by an immediate relationship such as a communication need or a known cabling solution. Technical relations can both be based on classification and immediate relations, strategic relations are always based on classification. Striving for modularization, the components with strong relationships should be grouped, and also components with similar properties should be grouped. Some of the information may not be known at the moment of analysis; these issues should be represented in such a way that the analysis is not influenced [Blackenfelt 2001].

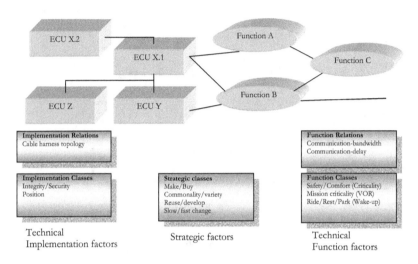

Figure 5-5 Factors for modularization

Technical function factors

One set of *functional* factors is the immediate *communication relationships* between the logical components. Functional relations are embodied by communication needs in the implemented network. The communication requirement can concern either the *bandwidth of the communication*, or the *delay in the communication*. A strong relationship is built by high bandwidth needs and requirements for low communication delays.

The other set of functional factors, the *functional classes*, concerns either the target or timing of the service as discussed in section 2.2. The target of service is either an individual in the case of personal services, or the driver in the case of more critical vehicle services related to the driving of the vehicle. Some services are personal services supporting the driver which make them partially vehicle services. Generally it is possible to distinguish comfort services that are less *safety-critical*, and safety services that are more critical and must be handled accordingly [Larses 2003a]. Separating safety-critical functions from non-safety critical functions through firewalls in physical, logical and temporal domains is a well known technique for computer systems [Watt 2000].

It is also possible to consider *mission-criticality* and the possibility for a vehicle off road (VOR) situation. To ensure a high availability of a system, a similar reasoning of introducing firewalls can be applied; systems that are critical for the availability for the system should be clustered and protected.

The timing of services can address services used while *riding, resting (but in the vehicle) and parking* as discussed in chapter 2. The argument for separating this kind of categories is a power management perspective.

Modern vehicles, and especially heavy vehicles, are becoming used for more and more applications outside the time span of actual transportation. Functions that are constantly activated should be clustered in order to reduce the part of the network that must be active to support this functionality; a proper clustering may also remove some of the needs for *wake-up* of ECUs.

Besides the three given timing categories, there is also a fourth that can be labeled workshop and includes diagnosis and maintenance work on the vehicle. This kind of services may be grouped with the driving category for modularity issues, as there should not be any power management issues in the workshop.

Strategic factors

Strategic factors are classifications of the logical components based on business issues that drive modularity. Four module drivers based on strategic factors are identified in line with previous work of Blackenfelt [2001]. These factors include the classification pairs of *make/buy*, *commonality/variety*, *reuse/develop* and *slow/fast change*. These four strategic pairs are considered important for automotive embedded control applications [Larses & Blackenfelt 2003].

The usage of strategic module drivers can be exemplified by the choice for a component to be out-sourced (buy) or developed in-house (make). When two functions both have the module driver "make" it could be argued that they strategically are similar and it would therefore be beneficial to place them within the same module. Also, it is here argued that it would be bad to place a "make" component together with a "buy" component within one module. Bought systems have to rely on standardized hardware interfaces and a well defined position in the hardware topology, while developed systems can be optimized for electric signals and software. The interfaces between the proprietary parts and the bought systems must be well understood, standardized and documented.

Considering the commonality/variety pair, the control units that are common in the system define the minimum platform and should be grouped if this means that some communication links and bridging control units are removed in the minimal system configuration. Some parts of the network can then be similar for all product variants and the technical content of this subset can be optimized accordingly. Handling variety can be contained to specific parts of the network.

Further, quick changes and the introduction of new or modified units are seen as a source of introducing faults and compatibility problems. This motivates the use of the strategic aspects "reuse vs. develop" and "slow change vs. quick change". The difference between these two aspects is found in the time perspective and knowledge about the future design. The reuse/develop aspect considers known changes up to the point in time that the architecture is designed for; if several parts of the system are about to be (re)developed the possibility to make a new module of these parts are greater whereas reused components should be handled separately. The slow/quick change aspect refers to the predicted change rate and future situation for a given component beyond the time scope of the architecture to be designed. The details of these changes are probably unknown. Components with a slow change rate can more easily be integrated in a module stable over time, while parts that are expected to change quickly should be excluded from such modules. The different strategic drivers are illustrated in Figure 5-6.

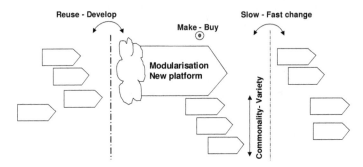

Figure 5-6 Strategic module drivers

Each component will be classified and attributed labels such as "reuse", "common" or possibly is no statement made for the module driver couple. Missing statements are ignored in the analysis and can be chosen afterwards to improve the strategic similarity. The results of the clustering analysis may indicate proper directions for some of the unknown pieces of information; for example, the choice of making or buying a specific unit may be influenced by the functional relations and the resulting position in the hardware topology. Also, the physical placement of a unit may be given by the communication needs.

Technical implementation factors

Technical factors concern issues in the implementation related to the hardware and physical interconnection in the wiring harness. The cabling should be determined by the cluster analysis proposed in this paper, but there may be some predefined cables existing due to the architecture legacy that must be accounted for in the analysis, based on the allocation of functionality in the implementation. The immediate implementation relations are given by the known *cable harness topology*.

There are two implementation classes that influence the modularization. First the *positioning* of a component is important. Components that are geographically remote in the system are more difficult to interconnect and the relationships between them should be minimized in order to minimize the need for cabling. The cost of interconnecting remote units includes both the immediate cost of increased cable length, and also the indirect cost of tracing cables through difficult sections. Difficult sections can be found in narrow passages as well as between parts that are moving independently of each other, typically the drivetrain and the chassis, and specifically for trucks: the cab and the chassis.

Further, some units pose a threat to the *integrity and security* of other units being possible entry points for interference and intrusion. Standard COTS units that are allowed access to the bus, and diagnosis interfaces are examples of possible hazards. Such hazardous units introduce negative relationships to, and should not be clustered with, safety-critical or mission-critical units. The possibility of interfering with such systems, either unintentionally or with malicious intentions must be minimized.

5.4 Architecture analysis with Keyfigures

Modularity is only one aspect of the EE architecture. Some issues that are not addressed in the modularization analysis can be captured and evaluated through the calculation of comparable keyfigures. Even though architectural decision-making is expected to be qualitative, it can be supported by quantitative methods with quantitative goals in terms of measurable qualities. A goal oriented approach with keyfigures is also suggested by the balanced scorecard [Kaplan & Norton 1996], and the use of such an approach will be further investigated in this section.

5.4.1 Applying the Balanced Scorecard for architecture design

Keyfigures are well known in business management applications. Traditionally, financial keyfigures and budgets have been used for corporate management. This approach may however fail to communicate the actual long term goals of management. A more goal oriented method is the balanced scorecard [Kaplan & Norton 1996] that extends financial keyfigures with keyfigures and goals for three other perspectives; the customer, internal business process and learning & growth, illustrated in Figure 5-7. This allows a possibility to introduce trade-offs and balance the influence of short-term (financial) and long-term perspectives. Applying a few proper metrics in line with the management approach of the balanced scorecard should be a valuable tool for the design of complex system architectures, like automotive embedded control systems [Larses & El-khoury 2004]. The keyfigures are able to communicate the product strategy across disciplines in the multi-disciplinary work and also allow sound system level optimization of the product. The system architecture trade-offs are documented and can be performed explicitly.

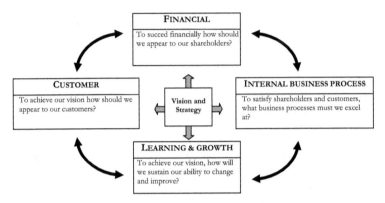

Figure 5-7 The four perspectives [Kaplan & Norton 1996]

Related to the balanced scorecard is a management work process with four parts. The original process activities include *translating the vision, communicating and linking, business planning* and finally *feedback and learning* [Kaplan & Norton 1996].

By defining a set of keyfigures, managers can build a consensus around the organization's strategy. By translating a vaguely stated vision to a set of objectives and measures, agreed upon by all senior executives, the expected long-term drivers of success are described. The articulated keyfigures can then be communicated across the organization and individual incentives tied to short-term financial goals can be avoided [Kaplan & Norton 1996].

The business planning process uses the balanced scorecard keyfigures as the basis for allocating resources and setting priorities, only those initiatives that aim toward the long-term strategic objectives are undertaken and coordinated. With the outcome of organization efforts measured by the balanced scorecard, a company can monitor short-term results and evaluate the strategy in the light of recent performance. The scorecard thus enables companies to modify strategies to reflect real-time learning [Kaplan & Norton 1996].

Through the keyfigures the strategy is communicated to the whole organization, not only the management level [Kaplan & Norton 1996]. It forces managers to explicitly make and communicate assumptions on the relationship between the chosen measures and the expected performance benefits. The goal orientation puts the strategy and vision, not control, in focus. This is equally important for the design of technical systems. A strong benefit of using explicit keyfigures is the documentation of the design rationale that is often hidden in the mind of the architect and usually not recognized. The design rationale is revealed and communicated to engineers in the organization.

The four steps of the work process can be immediately reused for architecture design. It is equally necessary for the architecture design process to perform these activities. The articulation of the vision lies in the design of the scorecard. The communication and linking is the communication of the chosen quality measures. With the keyfigures in mind, architecting can take place, creating resources (ECUs and cables) and allocating tasks to them (software and signals). The benefit of an engineering process is that the results can be immediately evaluated by performing a calculation on the proposed architecture. The final keyfigures can of course only be known after the architecture is finally implemented. In technical architecture design, indications of the system level impact of changes and alternative solutions are important. This suggest that trade-offs based on keyfigures would be useful also for technical work within a domain, or across multiple domains. Keyfigures are suggested as a mechanism that can solve communication problems across domains in a multidisciplinary design environment. They are also expected to be necessary for system analysis and optimization.

It is equally important to find the parameters that have leverage on the keyfigures in order to perform trade-offs. The importance of technical tradeoffs has been recognized earlier. In the architecture trade-off analysis method (ATAM) [Kazman et al 1998], a system parameter (i.e. properties and relations in a solution domain) that has significant influence on multiple requirements is referred to as a system trade-off point. For example, the number of servers in a client-server based computer system is a trade-off point since it promotes performance and availability but degrades security. The trade-off points become useful parameters that may be chosen as keyfigures as they influence the system architecture goals.

Although some properties are difficult to capture in metrics, it is possible to add soft measures for tradeoffs in line with the thoughts of the balanced scorecard. The interest in such metrics can also be recognized in efforts to achieve benchmarks for several soft values, like dependability of computer systems [DBench 2003; Kanoun, Madeira & Arlat 2002]. A benchmark must be uniform and repeatable independently of the measured system. However, for the purpose of architecture design, the selected keyfigures do not have to be generally applicable. It is enough that the measure can capture the strategic intention of the architecture design. Metrics are also seen as useful for evaluating software. Several metrics have been developed, and also theoretical frameworks for developing metrics exist [Talbi, Meyer & Stapf 2001].

Suggestion for perspectives of an Architecture Scorecard

It is explicitly suggested that companies implementing a Balanced Scorecard should adapt it to their specific context and needs. However, changes in the scorecard perspectives should be carefully considered in order not to include a too broad scope, and therefore lose the focus on the important issues [Kaplan & Norton 1997]. For technical systems it becomes even more important to restrict the number of keyfigures as there are many properties that can be measured.

Reusing the principle of the balanced scorecard in an architecture design context requires that a proper set of technical perspectives are established, a suggestion applicable for the design of automotive EE systems is illustrated in Figure 5-8. It is possible to use the original perspectives as an inspiration for developing a set of new perspectives applied to a technical setting. The financial perspective concerning resource consumption and efficiency relates to technical cost-efficiency, looking at resource consumption over a product lifecycle. The customer perspective reflects the externally perceived functionality, effectiveness and goal-attainment of the organization/system and thereby relates to performance. The internal business process relates to the systems ability to perform under any conditions and manage perturbations, reflected by the dependability and quality of a product. Finally, the learning and growth perspective relates to the long term sustainability of operations and can be seen as a reflection of the scalability and modularity of the architecture.

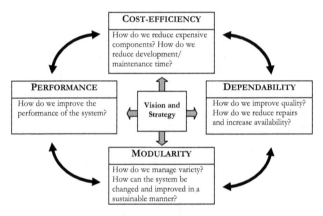

Figure 5-8 Perspectives of an architecture scorecard

The keyfigure approach has been evaluated in two Scania case studies, the PTA (partial truck architecture) case and the CTA (complete truck architecture) case. For the architecture design work a wide range of keyfigures was considered related to the four proposed perspectives. One example of a useful keyfigure is the number of *connection points*. Connections are a rich source of faults and failures in embedded automotive control systems. The positioning of control units and the allocation of IO ports are important design issues to reduce the number of connection points in difficult environments, these decisions also influence *cable length* which may be another useful keyfigure. The total number of connection points can only be reduced by utilizing mechatronic modules. Another relevant factor is the processing capacity of each control unit. The clustering analysis may suggest the merging of two units, but such cases must be verified for *processor* and *memory utilization* of hardware units. Even though the price of computing power is rapidly decreasing the performance of the hardware in the system may be lagging due to procurement issues. See Table 5-1 for a set of considered keyfigures.

Table 5-1 Considered Keyfigures

Number of connection points	Processor utilization	Number of mission critical connections
Connections in bad environment	Modularity	Number of part numbers
Cable length	Number of messages through gateway	Number of distributed functions
Number of cables in difficult passages	Gateway utilization	Number of widely distributed functions
Number of ECUs	Number of suppliers/ECU	Number of units developed in-house
Number of sensors	Number of suppliers/sensors	Number of pins/ECU
Weight	Component cost	
Bandwidth utilization	Number of Mission critical units	

Only a subset of the keyfigures was selected in each of the cases; this subset was expected to represent the system sufficiently well for the purpose of analysis. Other issues require a qualitative checking, for example the positioning of control units on a communication bus may introduce problems with *termination* of the bus. A CAN network should not be implemented as several interconnected star networks, but rather as a long bus with short stubs. However, in reality several units are often connected at electric centrals creating the unwanted star topologies. This may be acceptable as long as the cables are kept short, but the aspect must be considered in the analysis.

5.4.2 Partial truck architecture (PTA) case study experiences

In the PTA project, the architecture design problem was approached in a project of designing a partial EE network architecture at Scania. The project aimed at developing an improved solution for a bounded part of the EE architecture, mainly considering the hardware topology and cabling. It was assumed that the partial architecture would be clearly analyzable on its own. In the early stages of the PTA project, a set of keyfigures that reflect important quality attributes for system design benchmarking were discussed and established. The keyfigures included measures such as the total length of cabling, the number of connection points and the number of ECUs necessary for the implementation. The combination and weighting of the keyfigures was performed qualitatively in discussions. It was assumed that it would be impossible to distribute weights objectively to calculate a single benchmark comparison number. Benchmarking needed to be performed across the various variants/alternatives to be considered. Based on these assumptions a tool was developed to calculate the selected keyfigures in order to support the analysis of a given architecture design.

Initial modeling effort in the PTA case

In order to achieve a more structured evaluation of architecture alternatives a custom tool was built in Matlab and Excel. Commercial and academic tools such as Systemweaver from Systemite [2004], Ptolemy [Berkeley 2004],

Dome [Honeywell 2004] and GME (Generic Modeling Environment) [Vanderbilt 2004], could have been used for the design problem. However, the introduction of an integrating tool set requires changes in work processes and training. For this reason a more simple approach to system modeling was explored.

It was initially assumed that the project was small enough to be covered in a single model, and the initial analysis was performed in Microsoft Excel. The implementation utilized tables with objects and properties, matrices with relationships and properties, and algorithms calculating keyfigures. The modeling covered functional aspects and hardware/network architecture aspects, specifically modeling the allocation of functions to hardware, the network interconnection of hardware units as well as physical properties of the hardware. Excel was used to model the system, as it provided easy editing and viewing of information. Matlab was introduced as a complementary tool to perform some calculations too cumbersome for Excel to handle. The implemented tool calculates keyfigures describing the architecture, like cable length, number of connection points and number of conductors from cab to chassis. One set of keyfigures is based on information about a given architecture and given truck variant. The principle for the tool is illustrated in Figure 5-9.

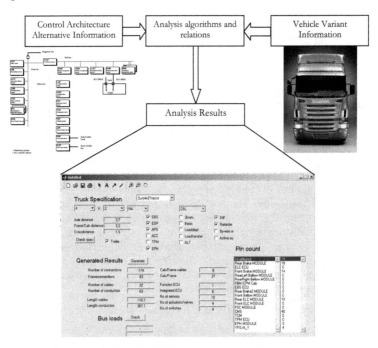

Figure 5-9 Initial analysis tool principle

For the PTA case analysis three different trucks were used. Analysis of several trucks and a comparison of the results captures the modularization aspect of the architecture to some extent. With proper modularization the keyfigures should scale according to the level of specification of the studied truck variant. A high specification truck should have more ECUs, longer cables and more connection point compared to a low specification truck. The three trucks were evaluated for a range of architecture alternatives. One alternative was designed to minimize cabling which was seen in low keyfigures for cable length. This alternative increased the number of connections of the frame as intelligence was distributed geographically in IO-nodes to reduce cabling. Other alternatives focused on the reduction of control units and connection points thereby increasing the need for cabling. The keyfigure tool provided useful data that supported the comparison of the different alternatives, and also supported the final decision making.

The method was quick and easy to implement but the problem of maintaining the model became more and more difficult as the project evolved and grew. Several conceptually different architectures were evaluated. In the end it was difficult to introduce larger conceptual changes and the tool was no longer helpful for general synthesis, but rather a case specific analysis tool. Further, there was no implemented support checking contradictions and validity of the models. No views were available for domain specific work, and changes in one domain required manual changes in several parts of the tool. System design was integrated with the domain specific design of functions, software and hardware architecture. At this point, one of the major benefits of model based development was lost, it was no longer possible to easily reuse the tool for similar cases.

The extended PTA case, the AIDA2 Tool

The AIDA2 model and toolset [El-khoury 2005] were inspired by previous experiences in developing modeling tools at the Mechatronics Lab at the Royal Institute of Technology. It has earlier been established that more than one model is needed in order to specify the control, software and hardware aspects of any mechatronic products, and models were developed to handle each of these aspects [Redell, El-khoury & Törngren 2004]. However, the integration of these models was generally performed in an ad-hoc fashion in order to deal with the current analysis issues at hand. No systematic approach was ever considered. One of the aims of the AIDA2 project is to provide a common and simple meta-meta-level modeling language that can be reused across many engineering domains, and to allow the easy integration of various models. A tool based on this language is expected to form a basis of communication between the various design teams.

The AIDA2 modeling approach is based on two simple concepts [El-khoury 2005]:

[1] A system can be viewed from different perspectives and any observer may only have a single point of view at any one time; each perspective

reveals a certain subset of the total properties of the system. Even though working with the same system towards the same goal, developers from the different domains use their own specific tools, providing their own specific views of the system to be developed. Each system view targets a specific audience, using that audience's familiar language (viewpoint), and concentrating on that audience's concerns [IEEE 2000].

[2] Each view can be decomposed in hierarchical abstraction levels. At each level, the system consists of a set of parts and their interactions; where the parts are considered systems in their own right.

The multiple views of the system can be interconnected across the hierarchies. Based on these interconnections, mechanisms were developed to allow a developer within a given domain to view the other aspects of the system from his/her own perspective, making view integration a good basis for information sharing. The proposed approach promotes the independent development of the views, allowing developers from each discipline to work concurrently, yet providing support for a holistic view [El-khoury 2005].

The AIDA2 model is at its very early conception stages and while a set of views for system functionality, software, computer hardware and geometry have been defined, much work remains before a stable modeling language emerges. For this reason, the PTA project provided a good cooperation opportunity between industry and academia. As a result, an early tool prototype based on the AIDA2 model that supports the design of functionality structure and cabling harness was developed in Dome [Honeywell 2004].

The experiences with using the AIDA2 tool for the PTA case was that it solved some of the previously introduced problems with the simpler application developed in Excel/Matlab. A clear improvement was the strict separation of functionality specification and hardware architecture views. The tool allows functions and the hardware architecture to be modeled on their own and then the views are associated for system level analysis. While AIDA2 initially focused on graphical representations, the benefit of entering data in another parallel format was acknowledged (For example, the tabular format of Excel). The graphical representation allows very easy navigation through the system, both within and between views. The meta-model brings available product data together into a virtual environment for analysis and allows reuse of product data throughout the entire product lifecycle, from development to production as well as maintenance. The main benefit of the AIDA2 tool is the validity checks of the system model. The tool both checks the validity of modeling decisions (by inhibiting illegal constructs) as well as the validity of analysis performed on the model (by identifying missing information). The benefits made it easier to work with the system and with system analysis. The cost for achieving these benefits was lifted from the system designer to the tool developer.

5.4.3 Complete truck architecture (CTA) case study experiences

In the second case study applying keyfigures the complete truck architecture was targeted. The second case utilized the proposed headings of the architecture scorecard, shown in Figure 5-8, as a basis for selection and design of keyfigures.

The headings of **dependability** and **cost-efficiency** were easily covered. For example, one important source of faults and failures in embedded automotive control systems is *connection points* that are easily counted and represented. The positioning of control units and the allocation of IO ports is an important design issue to reduce the number of connection points in difficult environments. The total number of connection points can only be reduced by utilizing mechatronic modules. Another reliability issue is to reduce the *number of cables* crossing between two independently moving objects, like the truck chassis and the cab. Further, *length of cables* and *number of components* are other easily analyzable figures related to cost.

Keyfigures for **modularity** were desired but it was difficult to easily extract useful measures for this purpose. The modularity required a more thorough analysis, which included a cluster analysis approach based on two modularization measures, described in the next section. Some modularity analysis was however performed through sets of keyfigures for different configurations. The possibility to scale the system for different product variants was evaluated and the minimum configuration was also analyzed. The **performance** was evaluated by the resource utilization of the system under given functional content. It is desired on one hand to have some reserve resources where new functionality can be allocated, but on the other hand also to achieve high utilization of resources in order to avoid the overhead cost from unnecessary bandwidth and computing power.

The improved architecture tool

In the PTA case it was possible to achieve system analysis results with a simple model. The results provided support for decision making related to the case and the development of the models provided insight in the system itself and the priorities of the design rationale. However, as the system grew in size, it became difficult to work with a single model. Further, it became very cumbersome and labor intensive to update the model in the later stages, but it was still possible to analyze the system variants for the given benchmarks. The AIDA2 tool added graphical navigation and validity checks of the model, both of these services were considered useful. However a simple tabular representation of the data was missing.

Building on the experiences of the PTA case, a tool that allowed keyfigures to be calculated for different product variants and different control system architectures was developed in a Microsoft environment using Visual basic and Access. The basic idea of the first keyfigure tools was reused. An analysis result requires a given architecture and a given vehicle variant. Architectures were defined by the allocation of functional modules to ECUs, and a certain topology interconnecting the ECUs. Variants were defined by a certain functional content. The implementation stored different architecture alternatives in separate Access databases while product variants were introduced as a specific table in each database.

In the tool it was possible to quickly try out alternative architectures and find the weaknesses and strengths of the alternatives illustrated by quantitative keyfigures. A screenshot of the main analysis window is provided in Figure 5-10. Modularity was analyzed through the method described in the next section, and to find architecture proposals, cluster analysis was performed.

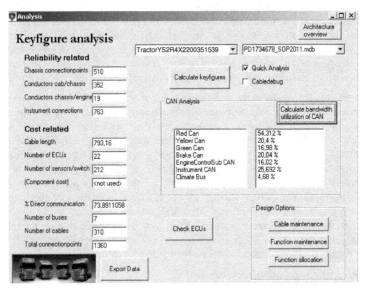

Figure 5-10 Screenshot of architecture scorecard tool

The new tool allowed much easier redesign of architecture alternatives through a simple in-tool user interface. While still providing quick and simple analysis the contents of the tool became much easier to maintain. The data in the tool was mainly based on product data available at Scania that was adapted to the tool format, but the design of the tool also allowed a rather straightforward addition of future functions using a simple model capturing assumptions about the function.

A problem encountered in the work was the lack of more detailed descriptions of functions and implementations in a machine readable format. Much of the documentation was not easily accessible by the tool and this may have reduced the precision in the results and also reduced the flexibility to easily redesign the architecture. However, based on the available data the tool still provided quick and useful answers to directed questions on how certain changes in the architecture would be reflected on system level. The keyfigure approach proved to be useful and beneficial for the architecture analysis work.

5.5 Architecture synthesis using the DSM and cluster analysis

In a hierarchical network, an important architectural design decision is the network topology. Assuming that the control system architecture is based on a set of control units with software and external interfaces interconnected on a bus, synthesis methods should indicate a proper allocation of control units on the bus, as well as allocation of external interfaces and software on control units. Which nodes should be placed in the same subnet, what external interfaces should the nodes have and what functions should be implemented in these nodes?

In this section a design method consisting of five steps that can be applied iteratively is proposed. The five steps are:

1. Data collection - Define logical building blocks
2. Data collection - Establish relationships between them (DSM)
3. Evaluation - Find an optimal clustering based on the relationships
4. Evaluation - Apply variations in cluster size and find the optimal clustering for each size
5. Interpretation - Interpret results qualitatively

The first two steps consist of data collection, while step 3 and 4 is an objective analytical treatment of data. In step 5 the results of the analysis is evaluated and interpreted in the light of the case. The method is further described below using examples from the ACC case study.

5.5.1 Data collection - Building the DSM

In order to modularize the product a model is needed that describes the products' elements and relations. In the proposed method a functional model, as described in [Pahl & Beitz 1996] where the relations could typically be information, energy and material transfer, is used. The functions could in theory be realized in any of the mechatronic domains. It should be noted that components such as the ECUs of the network in the current Scania architecture are not represented as elements, because they will rather carry the architecture and contain the functions and the relations. The functions should be grouped to modules, basically logical components, which then may be mapped to the hardware components.

The first step in the proposed method is to define the logical components of the system. The function blocks, or a set of logical control units, sensors and actuators based on the functional structure are used in the analysis. **The second step** is to define relationships between the logical components. In order to support the modularization the relations may be quantified based on the strength of relations [Pimmler & Eppinger 1994]. The factors, referred to as module drivers, elaborated in section 5.3, can be used as a template for the second step of the proposed method. For each module driver a set of quantified relations can be established. The module drivers are categorized as strategic and technical factors, where technical factors can concern the function itself or the implementation of it.

Functions and technical relations

Technical relations can be based both on the structure of the function and the implementation. The technical relations in the ACC case study only included functional relations and were based on the control design, Figure 5-11. The functional relations were associated with three levels of strength. Data that influence the control but are not a part of a closed loop were seen as the least strong relation. Closed control loop data relations were considered to be stronger. Closed loop control data with a high frequency were regarded as the strongest binding, while low frequency data was regarded an intermediate binding.

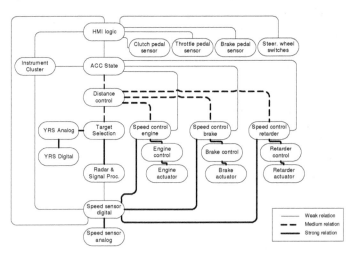

Figure 5-11 Technical relations based on control issues. The thickness of the lines indicates the strength of the technical relations.

Quantitative measures for the relations between the logical units can be established and represented in a design structure matrix (DSM). The possibilities of the DSM are discussed by Sharman and Yassine [2004]. The method used here is developed by Larses and Blackenfelt [2003]. In order to support the modularization the relations are quantified based on their strength. For example, the functional relationships shown in Figure 5-11 can

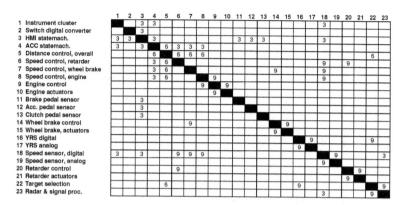

	1	2	3	4	5	6	7	8	9	10	11	12	13	14	15	16	17	18	19	20	21	22	23
1 Instrument cluster	■		3	3														3					
2 Switch digital converter		■	3																				
3 HMI statemach.	3	3	■	3							3	3	3					3					
4 ACC statemach.	3		3	■	6	3	3	3															
5 Distance control, overall				6	■	6	6	6														6	
6 Speed control, retarder				3	6	■												9		9			
7 Speed control, wheel brake				3	6		■							9				9					
8 Speed control, engine				3	6			■	9									9					
9 Engine control								9	■	9													
10 Engine actuators									9	■													
11 Brake pedal sensor			3								■												
12 Acc. pedal sensor			3									■											
13 Clutch pedal sensor			3										■										
14 Wheel brake control							9							■	9								
15 Wheel brake, actuators														9	■								
16 YRS digital																■	9						
17 YRS analog																9	■						
18 Speed sensor, digital	3		3			9	9	9										■	9				3
19 Speed sensor, analog																		9	■				
20 Retarder control						9														■	9		
21 Retarder actuators																				9	■		
22 Target selection					6																	■	9
23 Radar & signal proc.																		3				9	■

Figure 5-12 The technical relations in the ACC case represented in a DSM

be represented in matrix form in a DSM as shown in Figure 5-12. These relations form a basis for the further development towards the final DSM. For the technical relations the quantified strength of the relationships is based on the real-time control requirements as previously explained, this gives a reasonable indication of how the parts ought to be grouped. The three levels are identified with the values 3, 6 and 9 to indicate the differences in requirements. These steps are used rather than 1, 2 and 3 based on the assumption that the technical relations should weigh roughly as much as the strategic relations (which range from -8 to 8, further explained below). How the technical relations should be weighed against the strategic relations of course is a delicate issue, which is further discussed below. The direction of the real-time control communication is not regarded, this means that the DSM becomes symmetrical. It would be possible, and it is commonly exercised, to use the upper and lower part of the matrix to represent the direction of relationships, also enabling different strengths of the relationship to be recorded for each direction. In this case if the requirements of opposite directions differ, the toughest real-time requirement set the level. No negative relations are identified due to the nature of the studied product.

Strategic relations and constraints

The strategic relations between the functions are derived from module drivers or the strategic intention for the functions, which describe the purpose of the modularization. In the ACC case four couples of strategic intentions were used, explained in section 5.3.3: "make vs. buy", "reuse vs. develop", "slow change vs. quick change" and "commonality vs. variety".

Of course there may exist other strategic intentions but these eight (four couples) certainly are important in the ACC case. A part (logical component) in the product model will be attributed the label "reuse", "develop" or possibly no statement is made for the module driver couple. Thus, the part will relate to all other parts where a statement is done for the module driver couple. Similar statements will provide a positive relation and non-similar statements will provide a negative relation. Since most of the parts are likely to be classified it is probable that there will be a higher number of strategic relations than there will be technical relations.

The strategic relationships capture concerns with legacy and previous decisions. For example, it models issues such as the fact that GMS (gearbox management system) and EMS (engine management system) functions are traditionally developed in-house whereas BMS (brake management system) functions are mainly developed by suppliers. It also considers that there exists a number of engine variants whereas there only exists one retarder. Moreover, radar sensors and radar signal processing along with target selection are hardly technologies within Scania's core competence and thus will be bought. Table 5-2 exemplifies the classification of module drivers for a few logical components in the ACC case study. Some decisions about the module drivers are still not taken in the example, e.g. make/buy for the ACC state machine. The results of the cluster analysis based on the known module drivers may support later decisions. The open issues can be aligned to fit the properties of the designed architecture. The strategic relations are quantified using the same value range as Pimmler and Eppinger [1994], from -2 to +2, i.e. between a part which definitely should be out-sourced and a part that definitely needs to be developed in-house would lead to a "-2" relation. Since there are four module-driver couples the range will in total be -8 to +8.

Table 5-2 Module drivers for selected elements

Function	Reuse/Develop	Slow/Fast change	Make/Buy	Comm./Variety
ACC state machine	Develop	-	-	Comm.
Target selection	-	Fast	Buy	Comm.
Engine control	Reuse	Fast	Make	Variety
YRS, digital	Reuse	Slow	Buy	Comm

In a situation when a new function should be added to an available architecture a range of decisions are difficult to change and will remain fixed. For example, in the ACC case the control algorithms of the brakes will still be placed within the BMS and the brake actuators will still be controlled by the BMS. Since the ECUs are not represented in the product model the hard constraint is e.g.: the brake actuator and the actuator control should be kept together within the same module. Thus, on top of the four couples of strategic intentions, the constraints of the available architecture and legacy decisions are added.

The four strategic relation DSMs and the DSM representing functional relations are added together to form a combined DSM. This addition can be performed with a weight factor multiplied with each matrix, assessing the importance of each of the aspects. Such weights were not explicitly used in the ACC case, however the range of ±2 on each of the strategic relations and 0-9 on the technical relations provide a balancing of the importance of functional and technical relations as discussed previously.

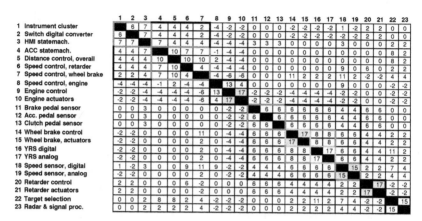

	1	2	3	4	5	6	7	8	9	10	11	12	13	14	15	16	17	18	19	20	21	22	23
1 Instrument cluster		6	7	4	4	4	2	-4	-2	-2	0	0	0	-2	-2	-2	-2	1	-2	2	2	0	0
2 Switch digital converter	6		7	4	4	4	2	-4	-2	-2	0	0	0	-2	-2	-2	-2	-2	-2	2	2	0	0
3 HMI statemach.	7	7		7	4	4	4	-4	-4	-4	3	3	3	0	0	0	0	3	0	0	0	2	2
4 ACC statemach.	4	4	7		10	7	7	-1	-4	-4	0	0	0	0	0	0	0	0	0	0	0	8	2
5 Distance control, overall	4	4	4	10		10	10	2	-4	-4	0	0	0	0	0	0	0	0	0	0	0	8	2
6 Speed control, retarder	4	4	4	7	10		4	-4	-4	-4	0	0	0	0	0	0	0	0	9	0	6	2	2
7 Speed control, wheel brake	2	2	4	7	10	4		-4	-6	-6	0	0	0	0	0	0	0	9	0	0	0	2	2
8 Speed control, engine	-4	-4	4	-1	2	-4	-4		13	4	0	0	0	0	0	0	0	9	0	0	0	-2	-2
9 Engine control	-2	-2	-4	-4	-4	-4	-6	13		17	-2	-2	-2	-4	-4	-4	-4	-2	-2	0	0	-2	-2
10 Engine actuators	-2	-2	-4	-4	-4	-4	-6	4	17		-2	-2	-2	-4	-4	-4	-4	-2	-2	0	0	-2	-2
11 Brake pedal sensor	0	0	3	0	0	0	0	0	-2	-2		6	6	6	6	6	6	4	4	6	6	0	0
12 Acc. pedal sensor	0	0	0	0	0	0	0	0	-2	-2	6		6	6	6	6	6	4	4	6	6	0	0
13 Clutch pedal sensor	0	0	3	0	0	0	0	0	-2	-2	6	6		6	6	6	6	4	4	6	6	0	0
14 Wheel brake control	-2	-2	0	0	0	0	11	0	-4	-4	6	6	6		17	8	8	6	6	4	4	2	2
15 Wheel brake, actuators	-2	-2	0	0	0	0	2	0	-4	-4	6	6	6	17		8	8	6	6	4	4	2	2
16 YRS digital	-2	-2	0	0	0	0	2	0	-4	-4	6	6	6	8	8		17	6	6	4	4	11	2
17 YRS analog	-2	-2	0	0	0	0	2	0	-4	-4	6	6	6	8	8	17		6	6	4	4	2	2
18 Speed sensor, digital	1	-2	3	0	0	9	11	9	-2	-2	4	4	4	6	6	6	6		15	2	2	7	4
19 Speed sensor, analog	-2	-2	0	0	0	0	2	0	-2	-2	4	4	4	6	6	6	6	15		2	2	4	4
20 Retarder control	2	2	0	0	0	6	-2	0	0	0	6	6	6	4	4	4	4	2	2		17	-2	-2
21 Retarder actuators	2	2	0	0	0	0	-2	0	0	0	6	6	6	4	4	4	4	2	2	17		-2	-2
22 Target selection	0	0	2	8	8	2	4	-2	-2	-2	0	0	0	2	2	11	2	7	4	-2	-2		15
23 Radar & signal proc.	0	0	2	2	2	2	4	-2	-2	-2	0	0	0	2	2	2	2	4	4	-2	-2	15	

Figure 5-13 A combined DSM with the sum of all technical and strategic relations included. The shaded green cells indicate the parts that must be kept together as a hard constraint.

The resulting combined DSM is shown in Figure 5-13 with the constraints represented by the shaded cells, this raw matrix is the basis for all analysis and evaluation to be performed. The constraints are not given any value since they should be seen as hard decisions, which cannot be negotiated.

Balancing the module drivers and other limiting factors in the implementation

The module drivers, summarized in Figure 5-14, can be quantitatively estimated providing a DSM for each factor. The factors can then be combined by summing the DSMs with weights assigned to each of the factors. The strength of a quantitative approach is that all small factors found to influence the architecture can be aggregated. This aggregation may not be perfectly weighted but it should be better than a qualitative aggregation where an individual must keep dozens of aspects in mind simultaneously. Also, a quantitative approach makes preferences explicit and allows an evaluation of several sets of weights, introducing a sensitivity analysis of the results.

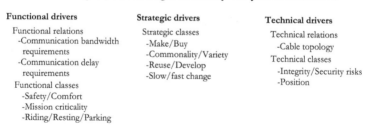

Functional drivers

Functional relations
-Communication bandwidth requirements
-Communication delay requirements
Functional classes
-Safety/Comfort
-Mission criticality
-Riding/Resting/Parking

Strategic drivers

Strategic classes
-Make/Buy
-Commonality/Variety
-Reuse/Develop
-Slow/fast change

Technical drivers

Technical relations
-Cable topology
Technical classes
-Integrity/Security risks
-Position

Figure 5-14 Summary of module drivers

Relating the module drivers

The proposed drivers have been used in the CTA architecture design case study at Scania. In this case study 32 logical units were defined and classified/related according to the proposed module drivers. Considering that some of the drivers are of a related nature it might be expected that it is possible to further simplify the module driver model by reducing the number of drivers. For this purpose an analysis of the cross correlations of the different drivers was performed based on the CTA data, the results are shown in Figure 5-15. By examining the cross-correlations among the proposed classification drivers it may be possible to find factors that are significantly correlated which can be interpreted as they measure the same thing. In the figure, drivers that are significantly correlated are indicated by stars ('*'). A single star indicates significance at the 95% level and a double star at the 99% level, the actual significance is also shown in the figure.

		Make_Specify_Buy	Platform_Option_Variety	Existing_New	Slow_fast_change	Driving_Stationary	Comfort_Safety	Criticality_VOR	Position	Security_Integrity
Make_Specify_Buy	Pearson Correlation	1	,353*	,554*	-,315	,264	,289	,356*	-,105	,384*
	Sig. (2-tailed)		,048	,001	,080	,144	,109	,046	,567	,030
	N		32	32	32	32	32	32	32	32
Platform_Option_Variety	Pearson Correlation	,353*	1	,281	-,002	,209	,331	,427*	-,172	,186
	Sig. (2-tailed)	,048		,120	,990	,251	,064	,015	,345	,309
	N	32	32	32	32	32	32	32	32	32
Existing_New	Pearson Correlation	,554*	,281	1	-,198	,130	,285	,382*	-,176	,129
	Sig. (2-tailed)	,001	,120		,279	,478	,114	,031	,335	,482
	N	32	32	32	32	32	32	32	32	32
Slow_fast_change	Pearson Correlation	-,315	-,002	-,198	1	,303	-,020	-,110	-,075	-,111
	Sig. (2-tailed)	,080	,990	,279		,092	,913	,547	,682	,544
	N	32	32	32	32	32	32	32	32	32
Driving_Stationary	Pearson Correlation	,264	,209	,130	,303	1	,448*	,331	-,314	,265
	Sig. (2-tailed)	,144	,251	,478	,092		,010	,064	,080	,143
	N	32	32	32	32	32	32	32	32	32
Comfort_Safety	Pearson Correlation	,289	,331	,285	-,020	,448*	1	,928**	-,515**	,421*
	Sig. (2-tailed)	,109	,064	,114	,913	,010		,000	,003	,016
	N	32	32	32	32	32	32	32	32	32
Criticality_VOR	Pearson Correlation	,356*	,427*	,382*	-,110	,331	,928**	1	-,495**	,471**
	Sig. (2-tailed)	,046	,015	,031	,547	,064	,000		,004	,006
	N	32	32	32	32	32	32	32	32	32
Position	Pearson Correlation	-,105	-,172	-,176	-,075	-,314	-,515**	-,495**	1	-,573**
	Sig. (2-tailed)	,567	,345	,335	,682	,080	,003	,004		,001
	N	32	32	32	32	32	32	32	32	32
Security_Integrity	Pearson Correlation	,384*	,186	,129	-,111	,265	,421*	,471**	-,573**	1
	Sig. (2-tailed)	,030	,309	,482	,544	,143	,016	,006	,001	
	N	32	32	32	32	32	32	32	32	32

*. Correlation is significant at the 0.05 level (2-tailed).
**. Correlation is significant at the 0.01 level (2-tailed).

Figure 5-15 Correlations among the module drivers

There are several significant correlations that can be seen in the figure. It can be concluded from visual inspection that there are two possible groups of related module drivers, [1] [Criticality_VOR, Comfort_Safety, Security_Integrity] and [2] [Make_Specify_Buy, Platform_Option_Variety, Existing_New], and also one completely uncorrelated factor [Slow_fast_change]. The driving/stationary factor seem to be somewhat correlated with one of the clusters [1] through the significant correlation to the Comfort_Safety driver, but it is mainly uncorrelated to the other factors. It is possible to look for principal components in the dataset and look at the eigenvalues of these principal components to establish the intrinsic dimension of the dataset [Bishop 1995]. An analysis performed on the eigenvalues of the covariance matrix of the data, using SPSS [SPSS 2005], verifies the general conclusion that there are three principal components (eigenvalues > 1), see Figure 5-16.

Component	Initial Eigenvalues		
	Total	% of Variance	Cumulative %
1	3,557	39,519	39,519
2	1,548	17,205	56,724
3	1,069	11,879	68,604
4	,822	9,131	77,734
5	,698	7,757	85,492
6	,617	6,853	92,344
7	,398	4,422	96,766
8	,242	2,687	99,454
9	,049	,546	100,000

Extraction Method: Principal Component Analysis.

Figure 5-16 Total Variance of extracted eigenvectors

The rotated principal components, shown in Figure 5-17, show the three principal component vectors. There is one distinct vector (providing 39,5% of the total variance in the complete data set) numbered 1 showing the correlation of the aspects of mission criticality and safety (*Criticality_VOR, Comfort_Safety*) associated to the placing of components on the drive train or in the cab (*Position*), and also to the risk of intrusion (*Security_Integrity*). Another vector (17,2% variance) numbered 2 is related to aspects of make/platform (*Make_Specify_Buy, Platform_Option_Variety, Existing_New*) buy/optional, and finally the expected change rate uncorrelated to these two clusters are found in the third component.

	Component		
	1	2	3
Make_Specify_Buy	,184	,801	-,178
Platform_Option_Variety	,140	,636	,267
Existing_New	,093	,781	-,102
Slow_fast_change	-,090	-,271	,850
Driving_Stationary	,339	,272	,670
Comfort_Safety	,766	,351	,230
Criticality_VOR	,754	,450	,110
Position	-,829	,036	-,124
Security_Integrity	,791	,099	-,131

Extraction Method: Principal Component Analysis.
Rotation Method: Varimax with Kaiser Normalization.

Figure 5-17 Rotated Component Matrix

Reflecting on these results it is possible to identify three main dimensions of arguments. The first dimension is related to the functional and technical aspects including mission criticality and position arguments. The second dimension is related to the strategic aspect of complexity management in the procurement process and variety management. The final dimension is another strategic aspect concerning the dimension of future change over time.

This suggest that the technical aspects drive the system essentially in the same direction and strategic aspects drive in a different direction and it is possible to emphasize either. However, a generalization to only two aspects may not provide the desired detail in reasoning. It is useful to maintain most of the variables to make more detailed decisions, for example a focus on criticality will separate systems with firewalls to create a dependable system, while a focus on position will reduce the need for cabling reducing cost and possibly also improving dependability as the cable harness is simplified.

5.5.2 Evaluating the DSM data

With an established DSM with quantified relationships between all components the evaluation phase to find an optimal clustering, **step 3**, can begin. The analysis of the raw data can be visualized by a rearrangement process. The matrix can be rearranged by renumbering the logical components and thereby changing their corresponding row and column in the matrix. In this rearrangement process blocks with closely related units, i.e. high positive relations, should be identified and placed in adjacent rows and columns. An illustration of a DSM with indicated clusters in adjacent rows and columns is shown in Figure 5-13. In the figure the clusters identified by the framed boxes contain strong positive relationships. The general principles of the method are further explained in Blackenfelt [2001].

Measures of modularity

When using the DSM for clustering parts to modules the idea is that strong and positive relations preferably should be kept within the module, whereas weak or negative relations should be kept between the modules. In order to describe how well a given set of modules satisfies these principles, metrics, such as Module Independence (MI) [Newcomb, Bras & Rosen 1996] and Average Ratio of Potential (ARP), may be introduced [Blackenfelt 2000].

$$MI = \sum_{mi=1}^{n} \frac{in_{mi}}{tot} \qquad (6)$$

$$ARP = \sum_{mi=1}^{n} \left(\frac{in_{mi}/pot_{mi}}{n} \right) \qquad (7)$$

The two metrics are shown in expressions (6) and (7). It is assumed that the components in the system are clustered in n modules, and the index mi in the expressions refers to a specific module. in is calculated by summing the relations within a given module, as indexed by mi. The in measure can be exemplified by the DSM in Figure 5-13. in can be calculated for the cluster indicated by the frame grouping logical components 8-10, concerning the engine control. Labeling the framed cluster as $mi = ec$, then $in_{ec} = (13+4)_{first\ row} + (13+17)_{second\ row} + (4+17)_{third\ row} = 68$.

tot is the sum of all the relations in the entire DSM and used for calculation of the MI-measure. *pot* is the potential maximum of a cluster, and is used for the calculation of the ARP. The value of *pot* is given by multiplying the number of relations within the module with a potential maximum relation score.

The value of *pot* in the ACC case example of Figure 5-13 is given by multiplying the number of relations within the module (i.e. 6) with a potential maximum relation score i.e. 17 (4x2 for strategic drivers and +9 for the functional control relation). This provides a $pot_{ec} = 17*6 = 102$. Considering the design of the ARP measure, the *pot* value normalizes the ARP measure to have a value between 1 and 0. The ratio of potential of the cluster indicated in the example becomes $68/102 \approx 0,667$. The ARP measure gives the average ratio of potential by summing the ratio of potential for each indicated cluster and then dividing the score with the number of clusters.

A higher value for MI or ARP suggests a better clustering. The main difference between the two metrics is that the ARP favors a higher number of smaller modules than the MI. In case there are only positive relations the MI would suggest one big module whereas the ARP differentiates between weak and strong relations. However, for a single element module the variable *in* is equal to zero, thus clustering of elements is still encouraged by the ARP measure. In **step 3** of the proposed modularization method an optimal clustering maximizing a selected modularity measure, should be performed.

Relational reasoning supported by quantitative clustering algorithms and metrics

In the search for good modules the solution space can become extremely large, as it is possible to vary both the number and size of clusters. It should also be noted that a DSM, as in Figure 5-13, represents only one possible arrangement of rows and columns. This makes it worthwhile to use some kind of clustering algorithms to support the search for modules. With a set of relation variables and a defined DSM with reasonable weights on relations, it is possible to treat the recorded values in a strict mathematical way. Using mathematics and selected measures, optimization algorithms can be utilized. One branch of optimization theory that considers similar problems is graph theory.

There are several different algorithms within graph theory that solve or approximate solutions to problems of clustering graphs with different constraints. Unfortunately, many of the problems are NP-complete. However, some algorithms are implemented and readily available in tools. In the ACC case a tool for graph partitioning called METIS [Karypis 2003] was used to establish an initial clustering. The METIS tool algorithm has some obvious differences to the problem we aim to solve. It does not take into account that the internal bindings of the cluster should be maximized and it also tries to build partitionings of equal sizes, which is not a constraint in our problem. The partitioning is still useful as an initial guess for further refinement. With a somewhat qualified initial guess it is easier to develop an algorithm that tries to improve the clustering, it was expected that less effort to ensure convergence of the algorithm was needed. A simple algorithm developed and implemented in Matlab for this purpose has been used in all the case studies for refining the results. The algorithm changes the clustering to improve the MI score of the system as much as possible in every step. Moving one or two components between clusters is considered as well as switching components between clusters. The limited implementation is simple and fast and can probably be improved to rigorously ensure optimal solutions. However, our effort does not aim to develop optimal algorithms to find proven optimal solutions, instead the results are only used as guidelines for further design. For this purpose the implemented algorithm is expected to be good enough, and the results are regarded as optimal even though this may not be true. In the end the results were inspected for improvements by hand.

An important thing to consider is the possibility to vary the size of the derived clusters. **Step 4** in the method prescribes that optimal clusterings should be found for different cluster sizes (again we regard optimal as the results of our simple algorithm). It can be noticed that the MI measure varies the cluster sized based on the average of the relations. A high number of positive relations increases the cluster sizes while negative relations reduce them. This makes it possible to create different solutions with varying cluster sizes by adding an *offset* to all of the relations. As the algorithm implemented in Matlab is based on the MI measure, high average scores make the clusters larger and a negative offset creates smaller clusters. It is possible to balance the DSM between positive and negative numbers using the offset. The choice of offset is a parameter that can be tuned to serve different purposes, for example finding clusters with different levels of strength. The different clusters derived by the described techniques were analyzed with the ARP and MI measures.

The infrastructure of the implemented Matlab scripts is shown in Figure 5-18. The DSM should be stored as a matrix in Matlab. A function is also available that can import matrix data from a file where numbers are stored in a tab-separated format. The MIopti subfunction provides an MI-optimal solution for a given matrix. The ARPopti function varies the offset to find a family of optimal solutions that covers the maximal ARP possible. The scripts deliver the results of the analysis in three text-files: *clusters.txt, ARP.txt* and *ARPclusters.txt*. The clusters file provides the clustering that achieves the highest ARP score. The ARPclusters file provides a matrix with the complete family of clusterings sorted with one cluster in each column. The ARP file complements the ARPclusters file by providing the ARP score for each of the clusters in the columns.

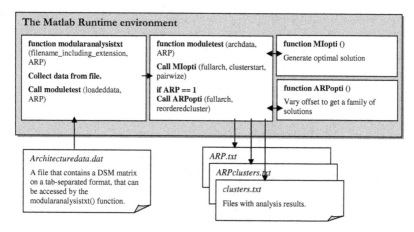

Figure 5-18 The Matlab cluster analysis architecture

5.5.3 Interpreting the results

The calculations and metrics do obviously not provide a final answer to the question of what architecture is optimal; the provided calculations must always be accompanied by qualitative reasoning and discussion. However, the qualitative reasoning may be supported by the results provided by these metrics. The method of cluster analysis for architecture synthesis is further explained with reference to the ACC (adaptive cruise control) case study detailed below.

The topology is expected to be a hierarchical network where control units will be assigned to different sub-buses. This assignment requires some analysis of how units should be clustered. But first functions must be grouped to modules, basically logical ECUs. The logical ECUs may then be mapped either to discrete components such as sensors and actuators, or to software implemented in the hardware ECUs. Assuming that the architecture should be based on control units in a network with sub-networks it is possible to assign the logical control units to different parts of the network according to the results of this clustering analysis. The system must be clustered at several levels of abstraction, the cluster analysis can be reused for functional grouping and software allocation to ECUs as well as for allocation of external interfaces to ECUs and ECU allocation to sub-networks.

Interpretation of the DSM cluster analysis in the adaptive cruise control (ACC) case

The ACC function was broken down into 23 function elements. Both technical and strategic relations were modeled following the ideas above. In Figure 5-19 a DSM with the sum of all relations in the cells is shown. Qualitatively, it is quickly seen that the engine functions including speed control of the engine (8) primarily have negative relations to most other functions and thus should be kept as a separate module. Speed control of the retarder (6) and the wheel brake (7) on the other hand fits well with the ACC distance control (5) and the state machine (4). The radar (23) including target selection (22) may be treated as a separate module although there are positive links to the ACC distance control and state machine. Qualitative reasoning like this may then be supported by the metrics such as ARP. For example, if the big cluster with functions 1-7 should be broken up, a better ARP is actually achieved by breaking it into 1-3 and 4-7, which makes sense, separating HMI functionality and core ACC functionality.

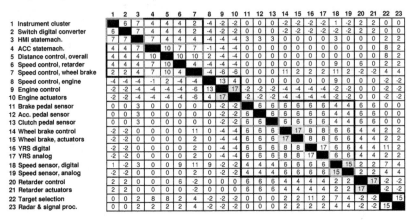

Figure 5-19 DSM matrix with the sum of all technical and strategic relations included. The different colored borders represent three clusterings. Black (ARP=0.66), Black/Red (ARP=0.66) and Black/Red/Blue (ARP=0.74).

Besides the combined matrix with both strategic and technical relations, shown in Figure 5-19, a matrix with only the strategic relations and a matrix only with technical relations were analyzed in the ACC case. These three basic DSMs have been analyzed with the automated clustering algorithm and with different offsets added. The combined DSM was also analyzed with a manual clustering based on the automated results. In the analysis the MI and ARP measures were calculated for each clustering.

The clusterings can be represented as shown in Figure 5-20. Each column represents a clustering where similar numbers and colors on the rows indicate logical components that belong to the same cluster. The first column in the figure is given by the clustering proposed by METIS. It can be noted that this initial clustering resembles some of the final solutions. It is also interesting to note how the results evolve with different levels of offset. The higher offset (moving to the right through the columns in the table), the more clusters are found, this is expected as the technical relations are only indicated with positive values. Low offset provides large clusters with weak grouping, and as the offset increases clusters become smaller but indicate stronger grouping. A small cluster found with a low offset applied suggests a strong module that should be well defined in the system; compare the reasoning about MI and ARP above.

The strong relationships indicate good modules and that the pieces of software should be placed on the same ECU. Intermediate strength relationships should be placed on the same bus and weak relationships can be placed anywhere in the network. For example, looking at the different columns of raw clusterings in Figure 5-20 it is possible to see that the engine speed control, engine control and engine actuator is a very strong cluster that should be maintained.

Component Offset:	METIS	Algorithm raw clusterings						Manual				Strat only raw				Tech only raw		
		0	2	3	4	5	6	Low	Med	High	Max	0	2	4	5	1	2	3
Instrument cluster	1	1	1	1	1	1	1	1	1	1	1	1	1	1	1	1	1	1
Steering wheel switches	1	1	1	1	1	1	1	1	1	1	1	1	1	1	1	1	10	10
HMI statemach.	1	1	1	1	1	1	1	1	1	1	1	1	1	1	12	1	1	1
ACC statemach.	1	1	1	2	2	2	2	1	2	2	1	1	1	1	13	2	2	2
Distance control, overall	2	1	1	2	2	2	2	1	2	2	1	1	1	2	2	2	2	2
Speed control, wheel brake	2	1	1	2	2	2	2	1	2	2	1	1	1	2	16	2	2	2
Speed control, retarder	6	1	1	2	2	2	2	1	2	2	1	1	1	0	0	2	2	6
Speed control, engine	3	3	3	3	3	3	3	3	3	3	3	3	3	3	15	2	2	3
Engine control	3	3	3	3	3	3	3	3	3	3	3	3	3	3	3	3	3	3
Engine actuators	3	3	3	3	3	3	3	3	3	3	3	3	3	3	3	3	3	3
Brake pedal sensor	4	4	4	4	4	4	4	4	4	4	4	4	4	4	4	10	11	11
Acc. pedal sensor	10	4	4	4	4	4	10	4	4	4	4	4	4	4	4	11	12	12
Clutch pedal sensor	7	4	4	4	4	4	10	4	4	4	4	4	4	4	4	12	13	13
Wheel brake control	4	4	4	4	4	4	4	4	8	8	8	4	4	4	4	8	8	8
Wheel brake, actuators	4	4	4	4	4	4	4	4	8	8	8	4	4	4	4	8	8	8
YRS digital	9	4	4	4	4	9	9	4	8	9	9	4	4	4	4	5	5	9
YRS analog	9	4	4	4	4	9	9	4	8	9	9	4	4	4	4	5	5	9
Speed sensor, digital	6	4	4	4	6	6	6	4	6	6	6	4	4	4	4	2	6	6
Speed sensor, analog	6	4	4	4	6	6	6	4	6	6	6	4	4	4	4	2	6	6
Retarder control	7	4	4	7	7	7	7	7	7	7	7	4	4	4	7	7	7	7
Retarder actuators	7	4	4	7	7	7	7	7	7	7	7	4	4	4	7	7	7	7
Target selection	5	1	5	5	5	5	5	5	5	5	5	1	5	5	5	5	5	5
Radar & signal proc.	5	1	5	5	5	5	5	5	5	5	5	1	5	5	5	5	5	5
ARP	57%	43%	58%	65%	68%	72%	70%	86%	66%	74%	76%	44%	57%	48%	33%	46%	39%	50%
ARPtot	6%	11%	9%	7%	7%	7%	7%	8%	7%	6%	6%	11%	9%	7%	4%	4%	3%	5%
MI pos	30%	81%	70%	57%	46%	36%	30%	64%	35%	30%	38%	82%	77%	67%	45%	31%	27%	23%
MI neg	0%	0%	0%	0%	0%	0%	0%	0%	0%	0%	0%	0%	0%	0%	0%	5%	5%	0%

Figure 5-20 Clusterings are represented in the columns. A specific cluster is identified by similar numbers on the rows and also by color coding. The cluster Manual (Low/Med/High) correspond to the cluster of Figure 5-19.

The different raw clusterings provided by the automated algorithm give good hints for the final manual clustering, all shown in Figure 5-20. It is noticeable that the clusters indicated by pure strategic and pure technical factors differ significantly. The technical relations cluster all the control software together and put the YRS (yaw rate sensor) and radar together, which suggests that the YRS in the BMS (brake management system) should not be used. The strategic relations make larger clusters, but interestingly enough some clusters coincide with the technical. The two aspects are also evaluated together and manually edited to solutions with high ARP scores. The ARP score can be increased through manipulation as the automated algorithm operates on the MI measure and not ARP. The clusters with high ARP seem reasonable, indicating that ARP is a useful metric.

Using both strategic and technical relations is highly useful as a guideline for system design in mechatronic systems. The freedom to reallocate functions provided by the use of software implies that more regard can be taken for strategic issues in future systems. The results show that strategic aspects together with ARP can give a relevant clustering of software based products. The exact values in the relations and the weights on strategic versus technical relations are however an open issue to be established for each specific case.

A DSM is very good to visualize the elements and their relations. It is also possible in a qualitative manner to get an idea of what happens when the module boundaries are moved slightly, which may be observed in Figure 5-19. The clustering algorithms using METIS and Matlab are however needed to support in the swapping of rows and columns in an effective manner. The clustering algorithms together with metrics such as ARP give hints to modular solutions, which then may be evaluated qualitatively by the user. Analysis of a set of clusterings with increasing offset makes it possible to order clusters according to module strength. In this process the ARP measure can be used for identifying clusters with a balanced set of elements, indicated by the ARP reaching a potential maximum. Using ordered clusters it becomes possible to assign specific elements to a chosen level of proximity. Two elements can be placed in the same ECU, on the same bus, or separated by a gateway in the network, depending on the strength of the cluster.

An important point is the need to analyze several functions simultaneously. In the ACC case the relations within a single function have been studied. With several functions using the same functional elements, as exemplified by the CTA case, it is very important to make a common analysis for the entire system, including all functions, to ensure that sub-optimization is avoided. This analysis will introduce more elements and also extend the number of relations as each function would correspond to a technical DSM. If technical relations are ambiguous or weak, strategic relations will play a more important role.

5.5.4 Case study CTA experiences.

The proposed method utilizing the DSM was also evaluated in the CTA case study. There a set of sub-DSMs were provided for a range of 10 different module drivers. Each of the separate sub-DSMs provided for each module driver were first normalized to have a value range between -1 and 1. The normalized factors were then weighted together by a multiplication factor between 0 and 10 to form a combined DSM. The combined DSM shown in Figure 5-21 was the resulting basis for the cluster analysis.

The same simple clustering algorithms used in the ACC case, implemented in Matlab, were used to support the search for modules through automated clustering. The different clusters derived by the described techniques were analyzed with the ARP and MI measures. By collecting data for each of the chosen aspects and weighting the separate DSMs together an assignment of logical control units to clusters representing network sub-buses can be performed. The results of a cluster analysis set are shown in Figure 5-22.

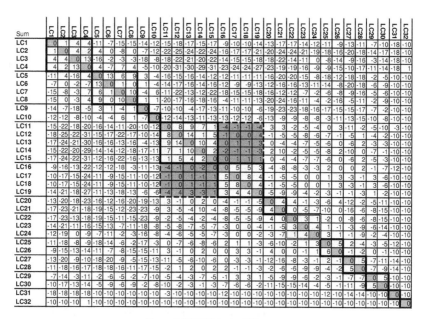

Figure 5-21 DSM in the CTA case

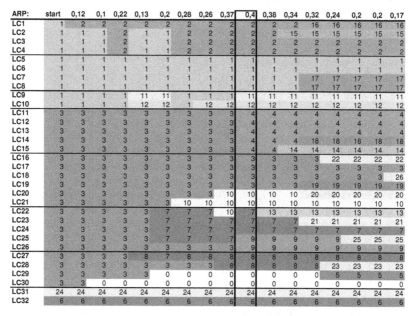

Figure 5-22 A set of clusters of logical control units with different cluster strengths

In the figure the logical control units (LC#) are represented in the rows. Each column indicates a clustering, where the number and color in the matrix indicate a specific cluster. From left to right the required strength of a cluster is changed through the addition of an offset in the analysis. The columns to the right show smaller clusters while the leftmost columns indicate only two major clusters. The highlighted column has the highest ARP value and would indicate a proper degree of separation. It is also interesting to notice that this column represents a major breakdown of cluster 3 into clusters 3 and 4. These results of the cluster analysis were fed back as new architecture alternatives for the keyfigure analysis.

By looking at one column at the time, and then at the sequential changes introduced by shifting columns to the left or right, it is possible to identify a hierarchy of modules. This hierarchy gives a good indication how sub-networks (indicated by the smaller clusters) should be integrated by backbones and core-buses in the network. Building hierarchies of modules using dendrograms has been proposed by Hölttä et al [2003]. What is interesting to notice in Figure 5-22 is that there is no strict hierarchy of clusters; for example, LC22 begins in cluster 3 changes to cluster 7, and then makes a short visit in cluster 10 before going back to cluster 7. Such irregularities exist due to the combined relationships in the cluster and are better accounted for in the type of diagram shown here than in a dendrogram.

Using the cluster analysis to capture design rationale

In the CTA case study the architecture alternatives were developed both by qualitative reasoning and the DSM cluster analysis. The weights for each of the relations were initially selected according to a general expectation of what a proper modularization strategy would be.

During the design process the weights were tuned and further developed. The architecture proposals synthesized with automated clustering were compared to qualitative designs and differences were analyzed. When the results coincide, the weights attributed to each of the aspects in the analysis are a good representation of the qualitative design rationale. With differences it is possible to modify the weights in order to *capture the rationale*, or to accept the proposed clustering and find flaws in the current qualitative reasoning. In the work with the CTA case study the general approach was to change weights to clarify the rationale. The final weights provided an articulated documentation of the actual architecture modularization strategy. However, there were some logical components that could not be re-clustered this way. These cases helped to point out important decision points in the architecture. In the case study it was found that functionality, criticality, position and to some extent platform and make/buy issues were prioritized in the cluster analysis.

Further, some logical components were sensitive to changes in the weights, the cluster analysis helped to understand the impact of changing the priorities and also helped clustering the sensitive LCs. The quantitative methods allowed useful *sensitivity analysis* of the results; it was possible to try a number of "what if" scenarios that changed the preconditions of the system architecture design.

5.5.5 Conclusions regarding cluster analysis

DSM and related measures are very good tools for analysis of mechatronic systems. The clusters found with high ARP scores in the ACC and CTA case studies do all more or less seem sensible. Earlier studies have suggested modularization based either on the properties of the product or on external requirements. This work shows how the two views can be combined as technical and strategic modularization factors. In a mechatronic system, including software, the strategic aspects may be stronger module drivers than technical concerns considering that software is more easily decomposed than mechanical components. Recognizing ARP as a useful measure it is shown how both technical and strategic aspects can be considered in a combined DSM matrix to create an ARP-strong clustering. Modularization based on the separate factors results in solutions with lower ARP values than using a combined DSM as shown in Figure 5-20.

As shown, mathematical methods and algorithms can be extremely useful to bound the solution space and support further qualitative reasoning. Furthermore, the introduction of an offset in the analysis allows reasoning about the strength of relationships and how to weigh technical and strategic aspects together. For good final results the technical and strategic aspects should be analyzed both individually and in combination and then be manually weaved, enabling cost-efficient solutions for the automotive industry. The actual assignment of weights to the different aspects constitutes an articulated modularization strategy.

5.6 A complete method for architecture engineering

Considering that keyfigures are useful for architecture and cluster analysis, and that the DSM is useful for synthesis of architecture alternatives, a combination of the two would provide a useful complete architecture engineering process. This combination was utilized and evaluated in the CTA case at Scania.

5.6.1 The CTA process

The architecture design process in the CTA case was based on three major components: The keyfigure tool, the DSM analysis and qualitative reasoning. In the keyfigure tool only analysis and evaluation of architectures were possible. The architecture alternatives were developed by qualitative reasoning supported by the DSM cluster analysis. The combined efforts showed the synergies of using a model based approach for multiple purposes. The same model used for the keyfigure analysis, stored in an Access database, also provided the basis for functional relations for the cluster analysis. The results of the cluster analysis could then be easily fed back into the keyfigure tool. A simple illustration of the design process is given in Figure 5-23.

By balancing the module drivers and also considering system issues outside the modular dimension, represented by keyfigures, it is possible to find a suitable clustering in the system architecture. The balancing was performed by tuning the weights for the DSMs of the different factors, and the previously described process of capturing the design rationale and performing sensitivity analysis by analyzing the weights was applied.

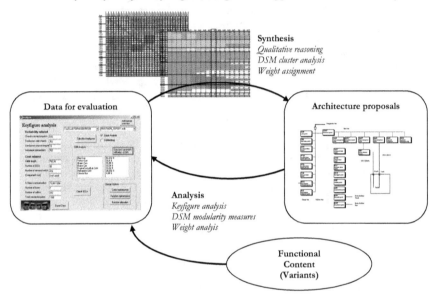

Figure 5-23 The quantitative tool supported design process

All the derived alternatives could then be compared by utilizing the keyfigures and modularity measures. Using the analysis tools to support the qualitative process helped to clarify design decisions and made the team aware of the design rationale. Documentation was easily created both for the design of the architecture and also about the rationale of the design, the three components were found to be useful and complementary for the work.

5.6.2 Problematic issues of quantitative architecture design

The quantitative approach proposed here, although shown very useful, has a strong limiting factor. This factor is the quality of the input data, as for any analysis method, the results are never better than the basic data used for input. A problem for extended analysis of embedded systems is the limited use of model based methods, specifically in the automotive sector. In order to fully reap the benefits of advanced analysis methods the quality of the input data, and the ease of preparing the data fit for use, must improve.

This does not imply that a single tool must be used for all engineering and product data management. Data exists dispersed and must do so due to the needs for domain specific tools in the development process, the data must however be integrated. The integration of data can be supported by an information model, further explored in chapter 6.

In the mechanical design of automotive systems, CAD models are used for reserving space for possible future changes in the design of the vehicle. Through these models it is possible to ensure that parts fit physically on the vehicle even if the design process is highly concurrent and distributed. The method is based on a proper modeling of the mechanical components in a CAD-environment. Such approaches, referred to as model based development (MBD), are becoming increasingly important also for the embedded control system as the number and combinations of functions implemented increases dramatically. However, the maturity of MBD in the automotive sector seems to be low, as discussed in section 4.3. Both functions and the actual implementation require more elaborate modeling support compared to current practice.

5.6.3 Conclusions

Decision-making at the conceptual level is expected to be qualitative but can be supported by quantitative mathematical methods. In the architecture studies described here the decision-making process was based on qualitative reasoning, but supported by mathematical methods. The proposed method suggests a combination of keyfigure evaluation and DSM analysis including automated clustering.

Achieving a modular architecture for embedded automotive control systems is highly beneficial for cost-efficiency and dependability. The architecture must consider the dimensions of functions, software, hardware and electrical signals. An operational system is developed by distributing efforts in all of these dimensions. Assuming that the functional dimension is given, and that the hardware topology is the most expensive to change, it is suggested to design this topology by performing cluster analysis in the functional domain based on proper factors. A range of factors that influence the modularity of the system is proposed.

The factors, referred to as module drivers, can be used for quantitative cluster analysis and are categorized as technical functional, strategic and technical implementation factors. By balancing the module drivers and also considering system issues outside the modular dimension it is possible to find a suitable clustering in the system architecture. Assuming that the architecture should be based on control units in a network with sub-networks it is possible to assign control units to different parts of the network according to the results of the clustering analysis.

The automotive industry requires that a system architecture is optimized not only for a single product, but for reuse over a range of products, and also for reuse over time with continuous improvements. To achieve these goals the product should be modular. Keyfigures have shown useful for managing trade-offs but do not easily describe modularity. To some extent, modularity over a product range can be evaluated by calculating keyfigures for a set of representative product variants. However, DSM and related measures are a more useful tool for evaluation of modularity; synthesis is also supported, based on the automated cluster analysis. If procurement and change aspects are included in the analysis, the quality of the modularization and the possibility to reuse the architecture over time is expected to improve.

A design process includes system analysis and synthesis. In the work described here the keyfigures were a valuable tool to support analysis and the DSM was a valuable tool to support both analysis and synthesis.

Further, the supporting mathematical methods clarified design decisions and made them more explicit. When the results of the quantitative analysis opposed the qualitative results the method helped to unveil hidden design rationale, or to correct faulty qualitative conclusions. Either the quantitative data was updated or the qualitative view changed. The method of using quantitative support for architecture design has proven very useful at architecture studies at Scania CV.

The current limiting factor of good analysis is the quality of the input data. The future success and elaboration of quantitative approaches for architecture design depends on the extent of model based development adopted. With an extended modeling of embedded control systems the architecting will become more engineering and less an art, increasing the usefulness of supporting mathematical methods.

6 Model Based Systems Engineering

Modeling has been discussed throughout this thesis and in this chapter the nature of the models themselves is further discussed. This is required in order to find a proper modeling technology that can be applied for automotive embedded systems. Information modeling is expected to be a precondition for a model based systems engineering process. In the end of this chapter some practical modeling cases, including the application of information models, are described and discussed in relation to the utilized modeling techniques.

Section 6.2 of this chapter is based on [Larses & El-khoury 2005a], 0 and 6.3 is based on [Larses & Chen 2003]. 6.5.1 is based on [Larses & El-khoury 2004] and 6.5.3 is based on [Larses & El-khoury 2005b]. 6.4 and 6.5.2 is written in collaboration with Ola Redell at KTH.

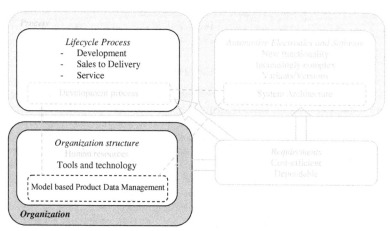

Figure 6-1 Chapter focus - Model based approaches

6.1 Meta-modeling and model frameworks

To provide a structure for the contents of this chapter it is useful to make some general points on the topic of models and abstractions. It is always possible to describe how something is described, thereby going to the meta-level. Wikipedia [2005] provides the following definition of Meta:

Meta *(Greek: "about," "beyond"), is a common English prefix, used to indicate a concept which is an abstraction from another concept, used to analyze the latter. For example "metaphysics" refers to things beyond physics, and "meta language" refers to a type of language or system which describes language.*

In UML 1.4 the Meta Object Facility (MOF) specifies four levels of abstractions as shown in Figure 6-2. This is to conform to the ISO and CDIF standard four layer meta-modeling architecture. In the MOF 2.0 specification [OMG 2003c] (and also in MOF 1.4) it is pointed out that four levels of abstraction is a rigid model and that it is possible to use MOF to specify any number of abstraction levels. However, for the purpose here, to structure a discussion of models for systems engineering the four layer model provides a useful picture.

| M3 The MOF Framework |
| M2 The UML meta-model UML |
| M1 Some UML models User model |
| M0 The real world User object |

Figure 6-2 The model abstraction levels in MOF 1.4

In the standard four layer meta-modeling architecture, shown in Figure 6-3, the *information* layer is comprised of the data we want to describe. The *model* layer is a structured selection of data (i.e. meta-data) provided in documents and models. The descriptions are used throughout the product lifecycle by different processes that have different requirements on the content and presentation of information. To maintain consistent and complete information, and to maintain traceability within and across the views of different users, it is important to have a core notion of the information required to describe a system. This can be achieved by the *meta-model* layer that defines the structure and semantics of models, while the syntax is provided by the model layer. Finally, the *meta-meta-model* layer defines the structure and semantics for the meta-models by providing a hardwired core [OMG 2002]. In this chapter the four levels will be referred to as *system*, *model (representation)*, *information model* and *framework*. These general labels for the four levels are also shown in Figure 6-3.

An information model is a semantic map where informational concepts and their relations are defined.

The information model can (and should) be developed based on a framework that besides providing a way to define meta-models also indicates different types of information, or system views, that are needed.

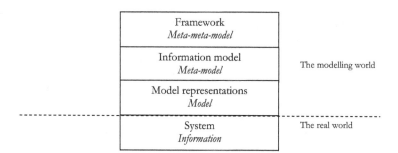

Figure 6-3 Generalized model abstraction levels

Frameworks providing a high level perspective on system views have been studied in the academic discipline of engineering design. Kruchten [2005] relates software design to the engineering design theories of Gero [1990], previously mentioned in section 6.3, and finds useful concepts: *"Having a broader understanding of other engineering disciplines--and software engineering's place among them--can benefit software engineers involved in multidisciplinary projects"*. Zimmerman [2005], also influenced by engineering design theories, investigates different approaches to information modeling of mechatronic systems. He proposes the use of the *Chromosome Model* [Andreasen 1992], based on the *Theory of Technical Systems* [Hubka & Eder 1988], as a meta-model guiding the design of the information model.

For product development, a process must also be applied on how the models should be developed and validated by the organization, besides the actual selection of models of the system. There exist frameworks that try to consider these aspects. A useful evaluation model for frameworks is described by Siegers [2005] who focuses on what he refers to as architecture frameworks. The evaluation model of Siegers proposes that architecting concerns:

Methodology – "how to" architect

Products – what models will be used to describe the architecture

Formats – the notation of the architecture product models (i.e. syntax)

Validation – ensuring concordance; assessing completeness and quality of the architectural solution

Collaboration – how the architecture team works together and with its stakeholders

Different frameworks emphasize different aspects and can be used to complement each other. For example, the DoDAF (Department of Defense Architecture Framework) provides a set of views that requires models, but no methodology guidance is provided. The TOGAF (The Open Group Architecture Framework) provides a detailed methodology based on nine process phases, but gives limited support in prescribing required work products and views. The Zachman framework provides a highly abstracted framework that classifies information into categories by using a 6x6 grid, but it does not provide any details on what models are required and what process to apply [Siegers 2005].

This chapter focuses on models and representations describing automotive embedded systems in terms of products and formats, but also gives a few pointers to issues of the lifecycle process concerning methodology and collaboration. First, to provide a background, some very general points of systems and system models at the framework level are discussed. Then a more defined framework will be described. Next some ways to provide information models and actual model representations are discussed. Experiences from case studies in the field are provided, and in the end an information model derived through the case studies is described. Some pointers to model representations implementing the information model are given in the case studies.

6.2　General System Models

To establish a framework at the most abstract level it is useful to look for some general principles for systems and modeling. The General System Theory (GST) aims to provide such principles and an applicable framework. System ideas have influenced several disciplines such as biology and economics, however, the focus in this section is on the study of systems as such. GST was established as a field of research in the 50's. The most commonly referred father of the theory is Ludwig von Bertalanffy. There is however a range of contemporary scientists who contributed in the field. GST has strong bonds to Cybernetics, Information theory and Control theory.

The more applied problem oriented systems thinking can be separated in hard and soft systems thinking. "Hard" methods include the field of systems engineering and assume that the system has a clear purpose and is optimized towards this purpose, they are goal-oriented. This assumption holds for human-made systems in general but breaks down for human activity systems, and also for some human-made systems where there exist conflicting goals and purposes. For these problems a "soft" methodology may be applicable. In this problem oriented approach developed by Peter Checkland [1999] model building (capturing abstract activities and issues) is in focus. The basic ideas of the soft system thinking are summarized by Flood [2000].

The hard systems thinking approach is applied in systems engineering and tries to arrange and describe the real world in a systemic manner in order to enable proper engineering. The exact details of the model describing the system are in focus. The first significant systems engineering was performed for telephone systems to ensure that all the different parts of the phone system interoperated reliably. The term systems engineering dates back to Bell Telephone Laboratories in the early 1940s, the first attempt to teach systems engineering as a subject was in 1950 at MIT. [Buede 2000] Today, systems engineering is promoted by the International Council on Systems Engineering (INCOSE) formed in 1990.

6.2.1 Concepts of GST

At the core of GST is the system definition, providing a meta-meta-model (framework) of the world. With a proper framework a foundation for knowledge creation is provided as models of the world then can be created and detailed. Unfortunately there is no common definition of systems and every other author (including the current) adds a new definition.

However, in the literature of GST there are some core ideas that are repeated and seem to be established in the theory. The ontology of any system theory contains three principal constituents: *unity*, *parts* and *relationships*. These ideas can be used as a core of a system model describing a system, enabling analysis and synthesis of systems. Further, it is commonly recognized that systems perform a transformation process and may have inputs and outputs. These concepts apply for both physical as well as for abstract systems. Also, each of the parts may be seen as a system of their own. The system concept can be applied recursively at any level of abstraction.

The Bertalanffy general systems model

The system view of the father of GST, Ludwig von Bertalanffy, focuses on the properties of the whole. Bertalanffy claims that there is no such thing as emergent properties of systems. The whole is more than the sum of its parts, but equal to the sum of its parts and the relations between the parts [Guberman 2002] "*If, however, we know the total of parts contained in a system and the relations between them, the behavior of the system may be derived from the behavior of the parts.*" [Bertalanffy 1969]

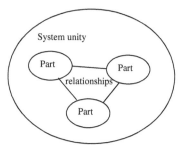

Figure 6-4 Ontological picture of system in Bertalaffys GST [Dubrovsky 2004]

It is possible to arrange the concepts relating to unity, parts and relationships according to Figure 6-4 in line with the ideas of von Bertalanffy. A well formulated and interesting criticism of his general systems model is given by Dubrovsky [2004]. Dubrovsky points out that the core of GST fails to formulate a single systems principle applicable to all systems, thus being general. The problems pointed out by Dubrovsky explain the lack of a structured core in current books on GST [Skyttner 2001]. He finds the origin of this problem both in the system concept of GST as well as in the related methodology applied. The criticism of the concept of system, as defined by Bertalanffy, is contained in two paradoxes, the paradox of emergence and the paradox of system environment.

The paradox of emergence relates to Unity represented as an entity of its own. However, if unity is the sum of the parts and their relationships then unity ceases to exist if any of the parts are removed. Thus unity is dependent on the parts and becomes a redundant concept. For the concept of unity to be justified, it must have properties of its own.

Further, if unity is seen as a system boundary then what is outside that boundary is called the system environment. However, as the system is interacting with the environment, the system and the environment must be two separate entities and thus the system is outside (not inside) the environment (*The paradox of system environment*). The paradox of system environment can be avoided by claiming that an organism is not a system but rather a "locus of behavior" [Dubrovsky 2004].

The Analysis-Synthesis system method

Kant provides an interpretation of unity that avoids the emergence paradox. Kant emphasizes the priority of unity over the relationships of parts. Unity is not emergent but exists prior to the relationships of parts. A system is not a matter of empirical observation, but rather a theoretical model or 'schema' determined by the combination of system principles and the subject matter. According to Shchedrovitsky [1966] (as referenced by Dubrovsky) the relationships are created in a process of synthesis, and the parts are defined in a process of analysis. A metaphor is provided by Dubrovsky [2004]:

Suppose one drops a teacup (unity), so it breaks ('analysis') into pieces (parts). One then glues the pieces together ('synthesis') in such a way that one can drink tea from it again (restored Unity). In this metaphor, the glue symbolizes a new addition (Relationship) that was not present in the teacup before it was broken, but had to be added in order to restore the cup.

Here it is important to notice that the results of the analysis are not only provided by properties inherent in the object that is studied (i.e. the cup) but also by the method of analysis (i.e. the way the cup is dropped). The provided results depend on how the beholder defines his parts, and relates them to the concept of unity. To provide another example, if the cup is only used as a counterweight on a set of scales the breaking of the cup is not an action of analysis. Then the unity of the cup is defined by the property of weight and not the form. However, weighing each of the pieces of the cup to find replacements for them would be analysis and synthesis to provide restored unity. This system definition, with unity as a complementary representation to parts and relations, is illustrated in Figure 6-5.

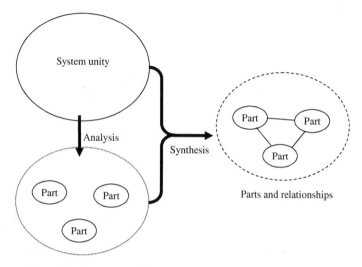

Figure 6-5 Logical relations among System constituents according to Kant [Dubrovsky 2004]

Further, three pairs of opposing properties are defined by Dubrovsky [2004] as form-content, complex-simple and external-internal. Based on these oppositions three system procedures are found. The first procedure concerns decomposition of an object into parts and is the opposite of composing. The second is the measuring of aspects of parts and wholes that is the opposite of configuration. The third is the insertion of an element into the object's structure, opposed by the extraction of an element out of the structure.

The goal orientation of systems

Bertalanffy claims that systems are teleological, meaning that they are goal-oriented and strive towards some end. The goal-orientation of systems has been criticized. Jordan, as referenced by Checkland [1999], distinguishes purposive and non-purposive systems. A mountain range is non-purposive while a road is defined as purposive. Checkland notices that the purpose is in the eye of the designer, builder and user of the road and not intrinsic in the road itself. He then proposes more useful distinctions that can be used to further clarify the goal orientation concept. He classifies five types of systems: Transcendental, Natural, Human activity, Designed physical and Designed abstract systems. Transcendental systems are beyond knowledge, unknown to man and can be ignored for our purposes. Natural systems can be analyzed, human activity systems can be analyzed and influenced, and designed systems can be analyzed and redesigned. A designed system has a function designed for a purpose. Checkland makes a clear distinction between activities (or systems) that simply serve a purpose, and activities (or systems) which are the result of a willed choice by human beings. Checkland chooses to label the first type serving a purpose as *purposive*, and the latter, when conscious human action is involved *purposeful*.

A summary of system concepts

Larses and El-khoury [2005a] provides a broader coverage of the general system theory. The concepts covered here are summarized in Table 6-1.

Table 6-1 General System Theory (GST) concepts

Concept	Opposite	Type	Description
System	-	Core concept	A system consists of a perceived unity, with parts and relations.
Unity	-	Core concept	Unity is one perception of a system.
Part	-	Core concept	A system can be decomposed into parts.
Relationship	-	Core concept	Parts can be related, related parts can form a unity.
Decomposition	Composition	Procedure	Definition of parts and relations from a unity. (Opposite: Building a unity with parts and relations.)
Identification	Configuration	Procedure	Measuring properties of parts and relations. (Opposite: Tuning the properties of parts and relations.)
Insertion	Extraction	Procedure	Adding parts to a system. (Opposite: Removing parts from a system.)
Teleologic	-	Goal-orientation	Systems that strive for a purpose
Purposive	Non-purposive	Goal-orientation	Systems serving a purpose (defined by humans)
Purposeful	-	Goal-orientation	Systems (designed by humans) serving a purpose (defined by humans)

Bertalanffy's GST model as interpreted by Dubrovsky is visualized in the class diagram shown in Figure 6-6a. In this model, the composition of Parts into a Unity is represented using the composition relation "contains". The Kantian view with unity as a schema for the decomposition into parts and relations is shown in Figure 6-6b. The difference lies in how the Unity and Parts are formed. The decomposition and relations are, according to Dubrovsky, inherent in the system in the Bertalanffy version, and formed by the beholder in the Kantian version.

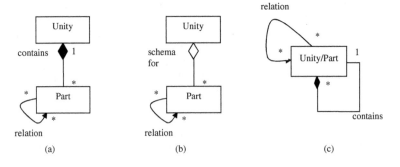

(a)   (b)   (c)

Figure 6-6 A UML class representation of the GST

These simple models do not, however, model the recursive definition of each part as a unit with its own decomposition into parts and relations. Figure 6-6c further develops the model, illustrating the recursive nature of the system definition. A recursive model is valid for both Bertalanffy's and Kant's system view, however only the Bertalanffy view (with the *contains* relation) is shown in the figure.

6.2.2 Weltanschauung

The notion of Weltanschauung (WA), originating from Kant [1790] and used in the general systems theory, provides a concept that defines the theoretical perspective used for analysis. WA is an image, a model, or an understanding of the world. In the context of system development, it refers to the perspective from which one observes and interprets a system.

A WA for system development can be characterized by the three fundamental questions: 'why', 'what', and 'how'. The 'why' is concerned with the intents and desired qualities of a system together with the concerned external conditions and restrictions. The 'what' is about the definition of solutions, while the 'how' is about the realization of solutions. Between them there is a natural precedence, that is, a 'why' is at a logically higher level than a set of possible 'whats', that in turn are at a logically higher level than a set of possible 'hows' [Checkland 1999].

In a large system, multiple WAs can be identified and arranged hierarchically according to the roles of the observers and the levels of abstraction that concern them, as illustrated in Figure 6-7. That is, the 'why' in each WA is related to the wider system; the 'what' corresponds to the system that is under consideration; and the 'how' is given by the underlying subsystems. Leveson [2000] building on Checkland [1999], refers to this as a "means-ends hierarchy" in the sense that the solutions at one level at any point in the hierarchy act as the goals (i.e. the ends or intents) with respect to the solutions at the next lower level (i.e. the means).

Why?
Increase house value

What?
Improve looks

How?
Paint house

Why?
Increase house value

Why?
Improve looks

What?
Paint house

How?
Brush paint house red

The estate owner
Investment system

The painter
Workplace system

The house owner wants to improve the value of his house. He/she finally decides to paint it. The task is given to the subsystem "paint" that performs the painting. The two systems have different answers to the questions 'why do it?' 'what to do?', and 'how to do it?'.

Figure 6-7 An example of Weltanschauung

Similar concepts in considerations of important issues in system development can be observed in many different approaches to system development and modeling proposed over the years. In the approaches to multi-viewed description of software systems, the WA is represented by a set of views as well as viewpoints [Kruchten 1995; IEEE 2000; INCOSE97]. In the Six View Organization of Models by Rechtin and Maier [1997], a set of system viewpoints[2] are identified using a range of questions that largely correspond to the why-what-how consideration.

What is interesting to notice is that in practice the focus is often placed on the 'what' and 'how', while the 'why' is expected to be captured completely in the top level system specification. For example, both UML and SDL support the description of system design, but require the users of the languages to keep the 'why' in their mind. A problem with such an approach is that the 'why' (i.e. the rationale) behind each design decisions is often omitted as the design goes on. As a consequence, eventual modifications or reuses can be difficult, for example these measures can easily lead to violations of required functionality, performance, or safety. One solution to the problem is proposed by Leveson [2000]. In the suggested software system specification language, each solution is connected to their corresponding 'why' or "intents" to the system specification by means of pointers.

In summary, a WA can indicate one of multiple *views*, as well as one of several levels of *abstraction*. This integrates two fundamental system mechanisms, *hierarchical decomposition* and *view based information filtering*.

6.3 The nature of systems

With a background on general systems theory it is possible to define a general systems model. A general systems model aims to provide a definition of the most abstract level as a meta-meta-model that defines how models of systems should be defined. Such a framework can be used both for managing system descriptions and traceability within a product, and also for management of concurrent product design and development by indicating the necessary documents and information content of a product specification. Hall and Fagen, as quoted by Weinberg, define system as: *A system is a set of objects together with relationships between the objects and their attributes* [Weinberg 2001]. In this thesis the mentioned definition of system is extended to be defined as:

A system is a bounded set of observed objects with relationships between the objects and their attributes, selected by an observer from a wider set of objects with relationships between the objects and their attributes.

---

[2] These viewpoints are: objective, form, function and behaviors, performance, and management.

This means that systems have an internal part and an external part and that a boundary exists, defined by observation, where the internal and external parts interact. It is also useful to define observer.

An observer is someone or something outside a system observing the system.

Further it is possible to clarify the relationships between system, purpose and observers.

A system may serve a purpose, providing a function in the eyes of an observer.

Based on this generic definition, we recognize some fundamental system aspects, including a system internal aspect, a system external aspect, a boundary, and stakeholders in the form of observers. See also Figure 6-8.

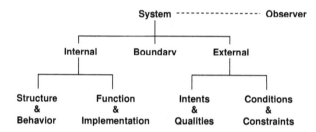

Figure 6-8 Characteristics of Systems

Internal aspect

The internal aspect of a system is concerned with things inside the system and the ways in which these things are brought together and managed to produce the whole. From an analytical point of view, a system consists of a set of parts and the associations between them. Here, the generic parts of a system are referred to as *system parts* and their associations as *relations*. These generic concepts can be applied in any domain of concern, for example, a system part can be a subsystem, a function, a software object, a processor, or a mechanical component.

Two important pairs of concepts are related to the internal model of a system, the *structure-behavior* and *function-implementation* concepts. These concepts are commonly mixed up and misinterpreted and an attempt to clarify the difference between the concepts will be developed below. The internal aspect of a system is characterized by the combinations of these two pairs, as illustrated in Figure 6-9. Combining the two creates four groups with the labels function-structure, function-behavior, implementation-structure and implementation-behavior.

A system can be considered with respect to its static structure or operational dynamics, which are often referred to as *structure* vs. *behavior,* see for example Rechtin & Maier [1997], Wieringa [1998]. While the structure represents "what a system is", the behavior corresponds to "what the system does". From a structure point of view, a system consists of a set of parts (e.g. people, machines, and communication buses) and the ways in which these parts are related to produce the structural whole (e.g. an organization structure, a machine structure, and a network topology). From a behavioral point of view, a system consists of a set of actions (e.g. initializing, processing, and communicating) and the ways in which these actions are related to produce the behavioral whole (e.g. a system mission in terms of a sequence of actions). The structure and behavior of a system are orthogonal and can be independently defined, however a defined structural part will have some kind of behavior. When a behavior is associated with a single part (or a group of static parts), it specifies what the part(s) does, for example how it can be started for execution: active, passive, by time, or by events.

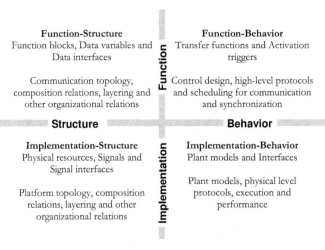

Figure 6-9 An internal view of systems

The second pair of concepts, *function* and *implementation*, deals with a system from different levels of abstraction. The function concerns a specification of relevant behavior from a specific viewpoint, concerning the system at abstraction levels that are independent of the realization means. The behavior specified as a function can be decomposed in a functional structure. The implementation focuses on the technologies and resources that create behavior for executing a given function specification.

It is not surprising that the closely related concepts are easily confused. A common mistake is to use the concepts of function and behavior interchangeably, thus discussing the function of the implementation and completely ignoring that functions can be arranged in a structure. A systems engineer once said, "*We talk about functions and software engineers talk about behavior*", which illustrates the mix-up of concepts. John Gero [1990] provides a better view on the topic with his function-behavior-structure framework. He provides a set of five elements that are labeled: function, expected behavior, behavior, structure and design description. Mapping the division of concepts proposed here into the Gero framework the function-structure maps to the function, the function-behavior to the expected behavior, the implementation-behavior to behavior and the implementation-structure to structure, the design description is not accounted for here as it is a concept related to the design process and not inherent in the system itself.

External aspect

The external aspect of a system is concerned with the things that are not controlled by the system, but still have significant influence on its existence and operation. In this thesis two domains of concerns are considered: *intents-qualities* and *conditions-restrictions*. These two domains of concern together define a problem-specific view of systems. The definitions of intents and qualities are largely in agreement with the traditional definitions of functional and nonfunctional requirements [see e.g. Kotonya & Sommerville 1998]. However, we have also classified the conditions and restrictions as first class concerns. This is based on the consideration that a requirement or quality specification is complete if and only if the assumptions (stated or not stated by the stakeholders) are also considered.

The domain of intents and qualities is concerned with the rationale behind the system. The rationale is not about the problems or solutions themselves, but the fundamental reasons behind them [Perry & Wolf 1992]. If the rationale is provided by a stakeholder it describes the objectives or ends that need to be achieved by a system. The intents and qualities of the external domain map to the function of the internal domain, the interfaces and variables that exist in the functional structure can be provided by the intents and qualities external to the system. The *intents* describe what actions the system should perform in terms of desired behaviors. For example, the primary intent of a truck is to transport goods from geographical location A to B. The *qualities* describe how effective the system should be when performing or realizing the intended actions, such as in terms of performance, reliability or delivery time.

The domain of external conditions and restrictions is concerned with the circumstances in which a system exists and operates in order to meet the given objectives. The conditions and constraints map to the implementation of the system, where conditions provide enabling inputs to the system and required outputs of the system, restrictions are bounds on the inputs and forbidden outputs of the system. The external *conditions* refer to the states and actions of external things upon which the fulfillment of the objectives depends. These can be physical conditions in terms of temperature and forces, informational conditions in terms of signals and data, or economical conditions in terms of market conditions. The external *restrictions* are the bounds or constraints on the external conditions such as in terms of physical laws, electro-magnetic interference, forbidden scenarios, accessible resources, etc. In different lifecycle stages of a system, the issues and details that are of interest in regard to these external things may differ. For example, an environment model that provides an abstract description of the actual external conditions and constraints is often used in early design stages. Extending the truck example with examples of conditions and constraints, it is possible to identify the road infrastructure as a condition and speed limits as constraints on the transportation system.

Boundaries

The boundary of a system is the border between its internal and external aspects. One boundary can be found in the implementation of the system, a different boundary is found in the function domain of the system. The implementation boundary interact with to the conditions and constraints provided by the environment of the implemented system. The function boundary is defined in relation to the qualities and intents of the system. See Figure 6-10.

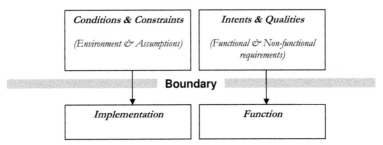

Figure 6-10 System boundaries

Observers as Stakeholders

An observer of a system may be a stakeholder. The stakeholders of a system refer to the persons or enterprises that have something to do with it. Their roles with respect to the system are often referred to as owner, customer, operator, developer, builder, and maintainer, etc. For example, the owner/customer influences the requirements as intents and qualities of a system; the user/operator chooses between alternative run-time behaviors via user interfaces; a designer may define the functional structure and behavior of a system; and a builder may affect the choices of implementation technologies. A stakeholder can be the society, an organization, a group of persons, or an individual. For example, the society can have requirements on the qualities of a product through laws and regulations issued by authorities.

Knowing the set of stakeholders and how their different views are related is an important issue for system development. It is necessary to ensure that the terminology and information content in the related views are consistent, and that the intents and qualities required by each stakeholder can be properly translated to a common set of requirements. The first step in this process is to identify all the stakeholders and understand how they observe the system.

6.3.1 A closer look at system design

With the proposed definition of system in mind it is possible to further develop an understanding of system design. System design is concerned with defining the constituents of a system, organizing their relations with respect to structures and behaviors, and planning for the implementation.

One necessary step to design a system is to define the system boundary and provide information on the system external issues. This is usually given in terms of a system requirement specification that states the system's intents and qualities, the external conditions and constraints, technology constraints and preferences, and the stakeholders' roles.

Since all these issues are concerned with the problems to be solved, they represent the problem domain of a system. In the design, the developers determine the system solutions based on the given specification, such as what parts should be included and connected and what actions should be performed and synchronized. The decisions may be based on methods from different engineering disciplines, domain standards and guidelines, heuristics, or consensus between stakeholders. The scope of design activities is illustrated in Figure 6-11. In the remaining part of this section some design related issues will be further discussed. Similar ideas are presented by Loureiro, Leany and Hodgson [2004] that propose modeling of requirements, functions and the physical system.

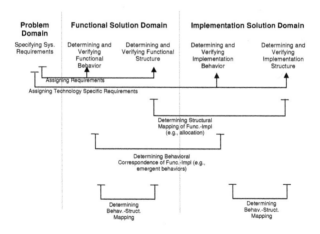

Figure 6-11 Scope of design activities

Relating the internal and external aspects of systems in design

The relationship between internal and external aspects of systems in system development can be illustrated as shown in Figure 6-12. The axes in the figure are based on the internal aspects of structure-behavior and function-implementation.

The horizontal axis in the figure represents a bounded set of behaviors. The bounds are given by constraints from natural laws. Further, in the behavior dimension there is a section of desired behavior; this represents the functional behavior of the system as defined by external intents and qualities. There is also a set of not allowed behavior representing external conditions and restrictions. This non-allowed behavior is influenced both by intents and also qualities in accordance with the external aspects of systems. For example, a not allowed behavior of a heavy truck is to drive faster than 90 km/h due to legal restrictions, for this reason there are technical systems limiting the speed. If the truck is used for racing this limitation is irrelevant and the boundary of not allowed behavior is moved.

The vertical axis shows the refinement of structure from functional to implementation concerns (0~100%). A specific implementation is defined by an implementation structure. The given implementation structure exhibits implementation behavior. The final product is reached with 100% implementation structure. The resulting implementation behavior is labeled emergent behavior and should cover the desired behavior if the implementation has been successful. It is also probable that the system shows other behavior, this is represented in the figure as redundant behavior. The verification of an implementation of a defined function requires that the behavior of the system can be specified as a function and measured in the implementation.

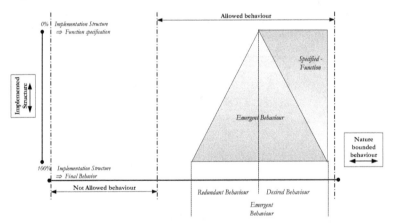

Figure 6-12 Structure, function and behavior

One important issue in system development is to balance between functional and implementation optimization by process measures. This has been underlined by Keutzer et al [2000]. They discuss how system design should be balanced between function and implementation optimization through the use of platforms. Platforms are seen as one way to allow parallel development of implementation and functions by using a fixed interface between function modeling and implementation.

Decomposition and composition

Decomposition and composition are related to creating a structure. Decomposition are used for *descriptive* reasons (analysis), labeling parts of a system in order to understand how the system works; composition is used for *constructive* reasons (synthesis) identifying necessary components of a system. In the following section way to perform decomposition are discussed, the rationale behind the alternatives can also be applied for composition.

There are two major categories of decomposition techniques: *behavioral decomposition* and *structural decomposition*. Again, the concept of behavior is commonly mixed with the concept of function. Behavioral and structural decomposition is often referred to as functional and structural decomposition. These two ways of decomposing a system have been discussed in problem solving strategies by Rasmussen [1993]. Further, Fischer et al [2000] notice that both of these decompositions are necessary in an object oriented model; in their report they show how decomposition is done in SDL-2000. The two dimensions of functional and structural decomposition are also underlined by Keutzer et al [2000].

Functional decomposition is applied to the functional behavior in order to create the functional structure [DeMarco 1978]. In behavior decomposition, the system constituents correspond to actions or functions/transformations. The decomposition is performed by dividing a behavior at a certain level into a set of behaviors at a lower level. In line with the previous identification of internal properties of systems, functions can be decomposed both according to behavior or structure. A behavior based decomposition of functions are represented by decomposed activities related by synchronizations, while a structural decomposition of functions would result in a dataflow representation showing the relations between the activities.

Implementation decomposition is usually related to a structural decomposition that corresponds to the boundaries between physical parts. The resulting implementation structure is the blueprint of the implementation. With a complete implementation structure it is possible to build the physical system. For implementation purposes the behavior is usually constructed by combining known implementation constituents. The implementation behavior is created by the aggregation of the parts forming the structural whole.

Over the years, other decomposition techniques have also been proposed, such as *event-based decomposition* [McMenamin & Palmer 1984], *device-oriented decomposition* [Yourdon 1993], and *subject domain-oriented decomposition* [Jackson 1983]. In the event-based decomposition, one system component (i.e. structural constituent) is defined for each stimulus from an external event creating a response in the system. In the device-oriented decomposition, one system component is defined for each external device to be controlled. In the subject domain-oriented decomposition, one system component is defined for each external entity about which data is to be stored or whose behavior must be controlled. All these approaches include functional and structural considerations.

Cross-domain mappings – assignment of functions

A mapping between function and implementation domains is necessary. One way to do this mapping is to find both the functional structure and the implementation structure and then relate the bounded artifacts to each other.

The ease of assignment depends on how the decomposition of functions and implementation has been performed. The behaviors in each domain must be aligned to fulfill the specification. The understanding and structuring of functional behaviors provides the first step in the development process. Functions define variables, transformations, states, communications and synchronizations at an abstraction level that is independent of the implementation. In order to determine how these issues will be implemented, one needs to determine the corresponding functional structure. For example, the functions of robot control systems can be layered into a hierarchy with mission control on the top and servo functions at the bottom [Albus & Proctor 1996]. This structuring constitutes the first step towards the implementation. The functional structure is mapped to the implementation structure under the restriction that the behavior of the implementation must comply with the specified behavior of the function.

Composition: System-of-Systems

A system can be one part of a larger system, and the constituents within a system can be systems on their own. A *system-of-systems* is a system built up from the integration of independently working sub-systems. Such a system differs from other systems in the sense that the (sub)systems are highly independent with respect to purposes, development, and operation. In Maier [1998], a taxonomy that distinguishes system-of-systems from other systems is proposed. In this definition, a system-of-systems is an assemblage of collaborative components that individually may be regarded as systems. These components possess two important properties: *operational independency* (i.e. the constituents can operate independently) and *managerial independency* (i.e. the constituents are separately developed). A system composed of complex subsystems that do not have both operational and managerial independence is not a system-of-systems, but a monolithic system. Depending on the existence of central control and explicit purpose of integration, a system-of-systems is also sub classified into different types: directed, collaborative, and virtual. In a directed system-of-systems, the systems work together to fulfill specific purposes under the control of a central management authority, such as in an integrated air defense system. In a collaborative system-of-systems, the systems must, more or less, voluntarily collaborate to fulfill the agreed upon central purposes, and the central management authority does not have coercive power to run the system, such as in the Internet. A virtual system lacks both a central management authority and centrally agreed upon purposes, such as national economies.

These distinctions are useful for reasoning about differences between systems regarding development and architecture. For example, in the development of a system-of-systems compared to monolithic systems, the focus is placed on the interfaces of subsystems and their integration. The developers often do not have full control over the subsystems and their properties. A related concept is *COTS-based systems* (CBS), denoting systems built up from commercial off-the-shelf products (COTS), often with the intention to quickly provide a service at low cost. Compared to system-of-systems, CBS are built on components with managerial independence, however the components may not be able to provide services unless they are a part of a greater system. A CBS can be a systems-of-systems but they do not have to be, they may be operationally independent. Similar to system-of-systems, the developers of CBS have limited control over the functionality, performance, preferred interoperation, and evolution of COTS constituents as the COTS are developed separately from the system.

6.3.2 An introduction to the Monty system meta-model

Here a meta-model for general systems based on the background on general systems theory and inspired by the ideas of Baron de Montesquieu is introduced. This meta-level model, hereafter referred to as the *Monty system meta-model*, aims to provide an easily understandable framework for identifying and relating various system views and required artifacts of modeling.

Montesquieu believed in democracy and identified three different powers that should be separated in the government of a healthy state: the "legislating", the "judging" and the "executing" power [Montesquieu 1748]. Combining the previously proposed system definition with the concept of Weltanschauung and adding some inspiration from the Montesquieu separation of concerns, the Monty system meta-model is achieved. The integration of these ways of understanding systems is illustrated in Figure 6-13. The model only supplies a general framework that through the use of other models can be applied both for technical systems as well as social systems. The Monty model is a meta-model capturing the spirit of systems rather than the systems themselves.

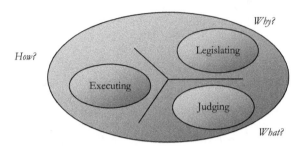

Figure 6-13 Relating Montesquieu and Weltanschauung

The Monty system meta-model identifies three fundamental parts of a system: *environment, function,* and *implementation.* These three parts play the roles of legislating, judging, and executing respectively for the entire system. The external purpose of the system, answering to 'why?', is contained in the *legislating* part labeled *environment.* The 'what?' correspond to the system internal function of the system and the *judging* part is thus labeled *function.* The actual things in the system *executing* the function are the constituents or subsystems that are labeled *implementation.* Between the three parts there are three defined interfaces, an *interpreting*, an *assigning* and a *resolving* interface.

Using the identification of separation in terms of Weltanschauung it is possible to understand more about the hierarchical relationship between the three parts. The logical precedence relation between the three Weltanschauung questions is also implied in the Monty system meta-model. The legislating role precedes the judging role, which in turn precedes the executing role. The parts, interfaces and their relations are further elaborated in this section. The set of concepts in the model are illustrated in Figure 6-14.

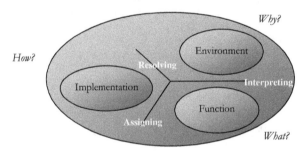

Figure 6-14 Concepts of the Monty System Meta-Model

Environment *– the combination of conditions or constraints that determine the existence, and consequences of a system.*

Environment. The environment constitutes the system's legislating basis and answers to the WA question 'why'. It contains the external concerns of a system in the domain of conditions and restrictions in which the system exists. These aspects would be captured in an environment model supplying the assumed conditions and restrictions in terms of the assumed structure and behavior of the physical environment of the system. The legislation of the environment is collected in a set of *environment objects*. The environment objects are a set of models of the environment that can be based on a variety of modeling techniques.

The intents and qualities are also external to the system. They relate to the environment but are rather captured in the function domain and through the interpreting interface as explained below.

Function – the functions and logics of a system that interpret what is required as intents and qualities and define a solution by structuring the functions.

Function. The function constitutes the system's judging basis and answers to the WA question 'what'. This means that the function part of the system connects to the external dimension of intents and qualities, thus interpreting the requirements posed on the systems from stakeholders. The function part can be extracted from a requirement specification.

The functions can be structured in a set of *functional blocks* providing the functional-structure. The selection of blocks and relations depend on the chosen method of decomposition (behavioral or structural) discussed previously.

Implementation – the collection of physical parts of a system that perform, carry out, or realize what is required of the system.

Implementation. The implementation constitutes the system's executing basis and answers to the WA question 'how'. It is the implementation means, such as the physical components, that are carrying out what is intended as prescribed by the functions and the description of their structure and behavior.

The implementation contains a set of *executing constituents* that represent the physical system boundaries as a group of implementation means, or physical parts. These (sub) systems can in turn be modeled as their own Monty system models. The minimum behavior of these system parts are thereby specified by the assigned functions. However, additional structure and behaviors may exist and may be allowed if these emergent properties will not lead to violations of constraints or patterns given in the corresponding function.

In a Monty meta-system, some of the dependencies between the constituents of the implementation are established via the assigning interface. In other words, the executing constituents can have both *functional* and *physical dependencies*. The physical structure must be defined and specified indicating physical relations between parts that can cope with the functional dependencies.

Interpreting interface – a selection of the environmental issues that are of particular concern in determining functions and behavior logics.

Interpreting interface. The interpreting interface exists between the environment model and the function of a system. It represents the selection of abstract environmental information that needs to be available for the functions.

This interface constitutes pure abstractions of the information available for decision-making and does not represent any physical sensor devices even if these sensors are the probable origin of the information. The interface can be modeled as variables or function blocks for information collection in the functions. By modeling an explicit interpreting interface the assumptions about the environment are made explicit. The information interface shows what assumptions of information availability that have been made for further modeling and design.

Assigning interface – the mapping of functional blocks to the implementation constituents.

Assigning interface. Each functional block, including representations of communication, of the function domain is mapped to a specific executing constituent in the implementation through the assigning interface. The assigning interface between the function and the implementation of a system extends requirements on the executing constituents, and maintains traceability by making relations explicit.

Within the assigning interface lies the implementation decision on what technologies that are to be utilized by the system. It also specifies how control is implemented, deciding on properties such as if control is to be distributed or centralized. The relations between the set of executing constituents is partly defined through the assigned functions. The interdependencies that are declared within the implementation must comply with the functional relations found in the allocation interface. An example of functional relations is control interdependencies that will require communication between the two executing constituents.

Resolving interface – constraints and physical interactions between the implementation constituents and the environment.

Resolving interface. The resolving interface exists between the implementation and the environment of a system and provides a means of characterizing the interactions between these two parts.

The resolving interface represents the physical interactions between the system execution and the real world, such as by exchanging forces, temperature, or other forms of energy. The interface is bidirectional showing both the influence of the environment on the execution parts and vice versa, representing for example sensing and actuating devices that constitute the implementation means for the environmental variables of functional solutions.

Hierarchies

A challenge for systems engineering is that different aspects of system decomposition and different levels of abstractions are often mixed and exchanged without a clear distinction. To overcome this problem, a clear description for the parts-whole and means-end hierarchies in a system as well as for their relations is necessary.

In the Monty system meta-model, the separation of the implementation into subsystems is given by the parts-whole hierarchy but the means-end hierarchy is also defined, since the function of each subsystem is given by the assigned functions of the current system. Thus, a Monty system meta-model is always applied to a specific system at a specific level of abstraction.

It is possible to change the level of abstraction by introducing more details about subsystems. Any system can be placed in a hierarchy of systems and subsystems. For this purpose, the executing constituent of a Monty system may in turn be defined as a Monty meta-system and receives its specification from the super system. This makes the Monty system meta-model recursive, the constituents of the implementation instantiate new system meta-models. A part of the functions of the super system is mapped to the implementation (or subsystem) where the function of the super system is integrated into the legislating environment of the subsystem. The rest of the legislating environment of the subsystem is built up by parallel systems on the same level of abstraction and also from the environment of the super system. It is also possible for similar subsystems to be instantiated indicating that redundancy is added. The reasoning of Monty hierarchies is illustrated in Figure 6-15.

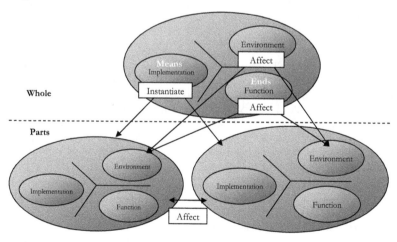

Figure 6-15 Hierarchies of Monty models

6.4 Required representations for automotive EE systems

Looking at the views required by automotive embedded systems one possible starting point is to utilize *the Monty model*, requiring proper modeling of the systems *function, implementation* and *environment*. Both function and implementation require a description of structure and behavior. A successful implementation provides at least the behavior required by the function specification, and no behavior forbidden by conditions and restrictions [Larses & Chen 2003]. This is similar to the Function-Behavior-Structure framework of Gero [1990], where functions are refined into behavior specifications that should match the behavior of the solution structure [Kruchten 2005].

The Monty model provides a useful framework for the information required for representation of a *single* system. However, to provide a useful framework for a *product range* of modular automotive systems the model must be extended with a representation of product *variants and versions*.

Further, the Monty model aims at representations of an actual system. In order to incorporate lifecycle process support, some representation of the verification and validation of requirements should be incorporated in the model. This can be achieved by the simple addition of a requirements perspective, representing the intents and qualities.

6.4.1 Technologies for automotive EE systems modeling

To support the entire lifecycle at system level the modeling must implement an information model that contains representations for the required domains presented above. The Monty system meta-model obviously only supplies the general frame. A meta model covering the different domains must be provided. In turn the meta-model must be represented by some modeling technique in order to be useful. Model constructs with a defined syntax appropriate for each domain are needed for implementing the semantics of the meta-model.

The main technology for developing information models is the standardized approach provided by STEP (ISO 10303). STEP is based on a powerful modeling language for defining information models called EXPRESS. Within STEP a set of standardized information models are supplied in the form of application protocols (AP) that provide models for different purposes. Zimmerman [2005] suggests that AP214, intended for automotive mechanical design processes, would also be a good basis for mechatronic systems.

Another technology is the UML language that is increasingly used for software modeling. The provided models can also be used for more general applications. Parts of the language are also useful for information modeling with examples included later in this chapter. The properties of STEP and UML have been compared. STEP is recognized to be a more powerful language for meta-models through the EXPRESS language, while UML provides better support for the actual model representations. STEP only provides simple support for text in ASCII format [Peak, Lubell, Srinivasan & Waterbury 2004]. A UML related approach is provided by the SysML project that increases the systems engineering awareness of UML-modeling through a variation of UML 2.0. The SysML standard includes an illustrative automotive modeling example in the specification that shows some system application of the language [SysML 2005].

A modeling approach, initially developed for aerospace applications, is provided by the society of automotive engineers (SAE). The SAE Architecture Analysis & Design Language (AADL) was published as standard AS-5506 in November 2004 [SAE 2004]. The AADL is a textual and graphical language that aims to support engineering of embedded real-time systems. A predecessor of AADL called MetaH was previously developed by Honeywell. MetaH, as well as AADL, is highly focused on the implementation of embedded systems in software and hardware and provides little support for broader systems engineering aspects.

Still another effort to provide a meta-model for automotive embedded systems is the ITEA-supported research project EAST EEA [EAST-EEA 2005] that proposes a language for automotive embedded electronic architectures, incorporating models of software and hardware. The EAST architecture description language (ADL) is defined using UML class diagrams.

The described approaches have different goals and scope, SysML is intended for supporting parts of the systems engineering process, EAST is a focused attempt to improve documentation of software and hardware for automotive embedded systems similar to AADL, while STEP concerns information modeling aiming at the proper design of information systems. SysML is intended to be aligned to the STEP standard, and specifically the application protocol AP233 for systems engineering. The intention is to map the SysML meta-model to the wider AP233 standard. Considering the three approaches in conjunction and comparing them it is possible to get a broader view on, and better understanding of, the modeling problem.

The environment of the EE system

In the Monty model the environment is a representation of the entities external to the system that either influences the system or with which the system is able to interact. For a model of a complete vehicle the environment includes entities on the road and in the traffic. For the embedded control system the environment includes not only the road and traffic but also the parts of the vehicle itself.

To be able to define functions, the available operands in the environment and their behavior must be known. Functions are defined by designing executions of transformations of the variables and events defined as a part of the environment. The environment relates to the operands in the process domain of the Chromosome model, where technical processes transform operands to produce effects, and a function is the ability to produce an effect [Andreasen 1992].

In SysML the need for an environment model is supported through a specific concept diagram proposed for modeling the operational context. The context diagram is a class diagram where objects that can interact with the system under design are defined.

EAST ADL lacks an explicit model of the system context. An implicit view of the environment is produced in a hardware system architecture model, but for an environment entity to be included in this model a solution of sensors and actuators must already be established. This reduces the benefits of the approach for conceptual development. AADL, also focusing on the implementation aspects, also lack an environment model in a similar way.

In the STEP approach the environment of the embedded system is provided by the integration of product data models for software as well as mechanical and electrical components. However, the complete system environment is not provided as an explicit model.

Function models

The functional abstraction is seen by Zimmerman [2005] as a key enabler for connecting technology specific parts. Functions are technology independent and bind heterogeneous disciplines together. The importance of implementation independent functions is also recognized in the Monty model [Larses & Chen 2003]. Function models are needed to allow the customer to explicitly specify the desired functional content of a product.

The design of functions to produce effects in the environment is theoretically well developed within control theory. The mathematical representation of a function as a transformation is a useful, implementation independent representation. Besides *dynamic functions* defined by control theoretic descriptions, there is also *static functions*, always provided by the product design exemplified by load carrying constructs like shelves and driver seats. Software based functions are always dynamic.

Functions should be defined without consideration of the implementation in order to facilitate configuration and tracing of the system, using the concept of function. A dynamic function can be modeled using data and control flow diagrams such as the activity diagram. SysML has a slightly different specification for the activity diagram compared to the UML specification and utilizes Enhanced Functional Flow Block Diagrams (EFFBD) [Bock 2003; Long 2002]. The EFFBD can be utilized to provide functional networks where the functional structure of the system can be optimized. By assigning the activities to components in the implementation, the communication between implemented components can be established.

In EAST, the abstract functions as defined by the Monty model are modeled in the vehicle feature model, while the models referred to as functions are closely related to the software implementation of the system. Behavior is recognized but not strongly supported in EAST ADL. State machines of UML 2.0 are proposed as behavior descriptions and a placeholder for external behavior models, such as Matlab Simulink is provided.

AADL does not have an explicit function concept.

In STEP, functions are represented as specific entities in the information model and related to product components. However, software components are not included in AP214 and modeling of behavior is not supported in the STEP standard.

Implementation models

The implementation of the system is often represented by domain specific models where details of the implementation are well specified. This includes CAD-drawings, software models, circuit board layouts and communication protocols. In the Chromosome model this is referred to as the assemblies and parts domain.

STEP, EAST ADL, AADL and SysML are all implementation focused, but the scopes of the implementation representations are very different. SysML is based on UML that is mainly intended for software systems, but tries to extend the modeling to cover also other aspects. Assembly diagrams are used for representation of physical components and parametric diagrams are used for representing physical relations between these components. The ability to express and analyze dynamic problems is however limited.

The EAST ADL covers software implementation and also some hardware aspects. There are four levels of modeling of the software and there are three levels of modeling of the hardware and one allocation view, the number of required representations relates to an implicit reference design process, regardless of the stated goal to be process neutral. A problem with the different representations is that information is provided redundantly and without stringent maintenance of related models, this may cause consistency problems.

AADL provides support for component based development and has proper models for both structure and behavior. Detailed models are provided for the electronics hardware platform as well as for the software.

In STEP the implementation and relations between implementation objects is represented with a high degree of details. STEP has originated from mechanical design and AP214 currently has no support for software, but it is general enough to be refined for mechatronic systems that include mechanics, electronics and software [Zimmerman 2005].

Requirements

Requirements are constructs used for the product development process. A requirement must be traceable from initial statement to implemented system. The requirements are the basis for verification and validation and must be included in the information model of the product. If requirements are represented as information objects, they can be attached to the information model used for the description and configuration of the final product. Weber and Weisbrod [2003], discuss the challenges for requirements engineering in the automotive sector. They conclude that requirements must be integrated in the information model to maintain traceability and the ability to manage change, this can easily be achieved in a STEP model.

When specifying a system, the environment, the functions and the implementation of the system must be properly modeled and linked [Zimmerman 2005]. In our view, the specification of functions captures functional requirements, while non-functional requirements are immediately related to the components in the implementation. Daniels and Bahill [2004] suggest a requirements process that combines traditional requirements, in the form of "shall" statements, with UML use cases, preferably extended with models such as activity diagrams, statecharts, sequence diagrams and collaboration diagrams. This process recognizes the need, and proposes a modeling solution for, separate functional and non-functional requirements; a separation well known in requirements engineering. SysML extends the support for requirements by providing a specific requirement diagram type. EAST ADL also have explicit support for requirements as an explicit artifact in the language. AADL does not supply an explicit way of treating requirements.

Representing variants and versions

Variants and versions can occur in all the dimensions of the product indicated by the Monty model. Variants in functional content exist due to different customer specifications; implementation alternatives are introduced due to technical considerations and cost-optimizations; changes in the environment can be induced by the introduction of new fields of application for the system. Further, requirements related to the different dimensions can vary due to different legislation in different countries, changes in market needs and technical innovations.

Users must be able to select the functional content of a vehicle and thereby create variants. To achieve a configurable product where the configuration is dealt with before deployment the variants and versions of the product must be explicitly documented and stated. STEP and AP214 provide good explicit support for configurations with a specific configuration object included in the information model.

EAST-ADL proposes a feature model where a selection of parameters that influence the configuration are declared in a flat list. Using variation points with conditions based on the provided parameters in the feature model, specific e-features can be provided [Freund et al 2004]. As a solution for variant handling for a product line, a representation of variations in three different artifacts is proposed. First a feature model that shows variation in functionality, then a product line architecture that shows variation in the physical architecture, and finally components and source code that actually implement the proposed architecture [Thiel & Hein 2002]. Utilizing the three artifacts, a tracing from feature through architecture to implementation is achieved.

SysML does not explicitly deal with versions and variants. Configuration is mainly discussed by referencing to mechanisms in the STEP standard, specifically in AP233. AADL provides some support for variants of how software is realized in hardware.

6.5 Case study experiences with model representations

To put all the proposed theories in action some case studies are needed. If the product data can be objectified it is possible to achieve lifecycle support where changes to the product structure can be contained in defined modules. STEP provides massive support for this purpose, but a simpler model can also be useful to maintain traceability and enable analysis of the system. A very simple model was utilized in the CTA case and provided sufficient support for the purpose, a simple model was also used in the SAINT case while the FDoc case developed a somewhat more extensive model. All these cases are accounted for and discussed here. In the case studies the topic of information modeling has been approached from the practical needs perceived in the process of engineering automotive embedded systems. The meta-models have not been based on any existing standard information models.

6.5.1 The CTA case

In the CTA case the goal was to develop a tool where the system architecture could be quickly analyzed, altered and analyzed again. The tool was based on a database and a simple model guided the development of this database. Of the system domains introduced in the previous section (6.4), only function, implementation and variants were included. The environment was not needed for this purpose and the requirements were captured in the actual analysis.

The core model of the database required a clear separation between function data and implementation data. Functions were modeled as function blocks linked by associated communication links, together forming a function network. The implementation was modeled as electronic units linked by associated cables, forming the network topology. The electronic units included sensors, actuators and electronic control units (ECU). The model allowed that function blocks were associated to electronic units and communication links were associated to one or more cables. These associations show the implementation of functions and communication links as software and signals in the physical system. Further, the core model indicated scenarios that contained a partial view of the functions through relations. A collection of scenarios could be related to a product individual indicating the functional content of this individual. See Figure 6-16 for an illustration of the meta-model of the product data.

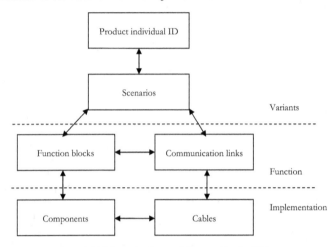

Figure 6-16 Schematic picture of data used in the CTA case

With a clear separation of function and implementation it becomes possible to trace a circle in the relations between the concerned modeling objects (Function blocks, Communication links, Components and Cables). This circle is clearly shown in Figure 6-16. The circle indicates that the objects are related as an equation system where it is possible to determine the fourth if the other three are known. This property was heavily utilized by the analysis tool in order to automate steps in the analysis.

A database was designed according to the information model. To provide data to the database several sources were used. An important source was a collection of UML message sequence charts (MSCs) documenting the CAN-communication in the system. Each MSC provides information about one scenario. The MSCs were processed for data regarding several of the information objects.

It should be noted that the extracted data actually represents the implemented software and not the abstract function, the reason for this is that the MSCs represent communicating software and the CAN-messages linking the software. For our purposes, lacking a software view, the product data found in the MSCs provided a useful substitute for genuine function documentation.

The MSCs, originally stored in the Telelogic Tau tool, were exported to a text format and then automatically imported into the Access database with a tool developed for the purpose. The data in the MSCs was stored as representations of electronic units, function blocks and communication links. The top-objects of the lifelines in the MSCs were stored as electronic units, the execution occurrence boxes on the lifelines were stored as function blocks and the signals in the diagram were stored as communication links. The remaining data conversion of properties of electronic units and CAN-signals was performed manually through cut and paste efforts.

Results

Collecting data and structuring the data according to the model two important capabilities were achieved. First, it was possible to automatically generate the relations between cables and communication links as the data relations formed a loop. This facilitated that new architecture solutions could quickly be evaluated as the allocation of CAN-communication was automated. Second, the link from product individual and the traceability maintained by the relations allowed quick reference to the product data and thereby easy and quick calculations of the keyfigures desired for the analysis based on a specific product individual. In total, the structured data management allowed fast results both for alternative product individuals, as well as alternative architectures. The analysis is further described in section 5.4.3.

6.5.2 The SAINT case

In the model truck project SAINT the focus was placed on achieving lifecycle support for flexible function development and configuration. In an early stage, three different types of configurations were identified and addressed: *Production time configuration* refers to the creation of a product variant through a selection and composition of hardware and software components, and through the setting of configuration parameters. This configuration is based on a customer specification. *After market configuration* refers to a re-configuration in the customer support process and is performed in a service station after delivery. For this type of configuration it is preferable to only use parameters to avoid logistics and compatibility problems. *Run time configuration* is the automatic adaptation of a truck when external systems (such as an advanced trailer) are connected or disconnected. This supplies the most flexible solution but also introduces other issues that we do not cover here.

The information model

Based on the requirements for configurability an information model was developed. The information model utilized in the SAINT project, shown in Figure 6-17 with UML notation, is based on the ideas in the Monty model and entails separation of function and implementation, and also separation of structure and behavior. The information model was realized in a PDM system which facilitates the navigation among modeling elements during development and later configuration, and helps in defining and enforcing the rules governing the feasible product variants. Each of the objects in the information model was implemented as *business objects* in the chosen PDM solution Matrix10 [MatrixOne 2005].

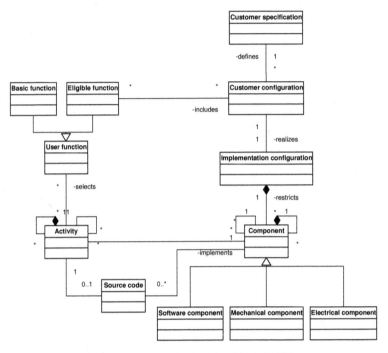

Figure 6-17 The Information Model used in the SAINT case

The model uses activity diagrams (EFFBD) from SysML as a means to model and integrate user functions. The *activities* themselves carry partial behavior and are explicitly shown in the information model. The activities are structured to functions in activity diagrams, and the behavior of functions is thereby defined. The activity diagrams are represented as *user functions* in the information model. The complete vehicle system can be shown in a single activity diagram, and subsets of this activity diagram define the different user functions. User functions were classified either as *basic functions* that always were included in the vehicle, or *eligible functions* that was optional for the customer to specify. The *customer configuration* object was linked to these eligible functions and was derived from the *customer specification* object.

Implementation *components* have behavior that must match the required behavior from one or more activities. The structure of the assembled components is defined in the *implementation configuration* object. The behavior of the assembled system must match the set of functions required by the customer. It is interesting to notice that the activities initially were intended to map immediately to the components, and this is also the case for non-software components. For software components it was found beneficial to map the activities to source code objects that in turn were related to the actual binaries running in the processors. The software was more easily structured in terms of source code, but it was also very easy to establish the mapping from a set of source code objects to the actual binary.

The product data model used in the project supports both the development and the production time configuration processes. For each information object in the product data model, steps in the development process that establish such objects must be introduced. During the design phase, *requirement objects* were defined and linked to functions, activities and components. These objects in the product data model are not shown in Figure 6-17 but exist in the implementation in the PDM system.

The SAINT model is very similar to the model used in the CTA case with a few major differences. The product individual is captured in the customer specification object that relates to a customer configuration object that is connected both to functions and immediately to the implementation. In the CTA case the product individual only related to the functions through the scenarios. Further, the loop of functions, communication links, cables and components is not included in the SAINT model. Neither the communication links, nor the cables are included in the model. These objects are not important in the SAINT case as the cabling is given and no analysis of the communication needs was intended.

Representing functions

In the SAINT project we separate the specification of logical functions from their implementations in software and electrical and mechanical hardware; thereby facilitating configuration through the concept of function. User functions can be individually selected by the customer of the system and they are also used for navigation in the diagnosis tools.

We found SysML to be the most useful source of inspiration for the purposes of the SAINT project. SysML provided us with good representations of functions through the *activity diagram*, but required that configuration mechanisms were tailored as needed. We found the EAST [EAST-EEA 2005] model too limited, and STEP to be too focused on the information modeling (which is important for the IT support infrastructure but helps very little in the actual representation of the different information objects). Our proposition is to utilize activity diagrams to describe the function architecture and then perform configuration management on leaf activity level.

Functions must be described with both structure and behavior. The behavior is defined in the leaf activities while the structure is defined by the composition of the activity diagrams. To further improve the model description of functions, sequence charts can be added that illustrate the actual behavior of each user function. The activity diagram provides a good overview for the function architecture while the sequence charts are good for function design and provide an intermediate step towards implementation.

Functions are linked to the implementation through the leaf function blocks in the activity diagrams that are related to components in the implementation model. A component may be an electrical, software or mechanical component, in effect enabling the specification of any implementation of the function. The activities carry the actual behavior description of the function that should be matched by the components implementing the activity. It is interesting to notice that the software is represented as source code for the purpose of facilitating configuration.

External tool support

The data indicated by the information model is used and developed in domain specific tools that store data in their own formats. To avoid inconsistencies in product data it is recommended that a common central database is utilized, as further discussed in section 4.4.2. This can be implemented in a Product Data Management (PDM) system. A PDM system is easily integrated with other tools and information that is relevant across multiple tools can be extracted and synchronized.

The domain specific tools utilized, including UML tools, should be integrated in the PDM system through provided API's. In our case study much of the work with the PDM system was performed manually due to the limited amount of time available for the project. However, the service application extracting software binaries and parameter sets that can be loaded into the ECUs was written in Java and interacts immediately with the API of the PDM system.

When the system development was finalized, or at least has led to a first release, the product data model had also been instantiated with model entities that were structured and easily navigable using the PDM tool. At this point, the model information was ready to be configured into product variants. A production time configuration is initialized by a customer making a selection of functional content and non functional properties of the desired product variant. This selection is performed through the service tool. The choices are stored in a *Customer configuration* that includes a list of chosen eligible user functions, and selected properties such as engine power and cab color.

The provided selections are traced through functions to components and occasionally immediately to components. If the customer configuration has provided enough and valid information, a complete *Implementation configuration* is created. Since activities may be implemented in many different ways, by different components, a set of simple configuration rules needs to be applied to distinctly relate an activity to a single component for a given product variant. In the end an implementation configuration that uniquely defines a product variant by listing a set of components and corresponding parameters can be generated.

The constraint mechanisms that enable variant handling are implicit in the information model: For example, some user functions are mutually exclusive (Standard cruise control and Adaptive cruise control) and hence the selection of one of these restricts the selection of the other. Further, while each user function unambiguously defines one or more required activities (adding no configuration freedom), an activity may be implemented in a number of different ways by different components. As a result, an important part of the configuration process is to identify a valid set of components that implements all activities defined through the selection of user functions. The conditions and strategy related to this identification must be supported, and is provided in the service application. A specific condition is that a *Customer configuration* may explicitly specify some components.

A platform for flexible systems

To support flexibility in the configuration and reconfiguration of the truck control system, a platform for component based software development is needed. In SAINT, a software platform was developed that provides the application developers with a standardized interface for communication and connection to external hardware units, independent of the final allocation of the software component. The platform is based on the OSE Epsilon operating system which gives good support for transparent communication, irrespective of process allocation. As a result, mapping of the various software configurations, specified and generated from the configuration process in the PDM system, to the corresponding hardware platform is supported.

The flexibility of the platform is also manifested in the support for "plug-and-drive" functionality, which permits the system to reconfigure on-line when more or less advanced external devices are connected. For example, the system was prepared for two different types of trailers, where the more advanced trailer was equipped with a local network of three micro controllers. These micro-controllers handle functions such as rear-wheel steering and distributed control of braking. The truck control system adapts depending on the type of external trailer connected. The control system either takes complete control over the trailer or, in the advanced trailer case, distributes some of the logic to the local micro controllers.

Results

The SAINT case concerned modeling to support configuration management of automotive systems, it was concluded that an information model supported by a meta-level framework is needed for traceability and configuration management.

A proper separation of function and implementation in a way that does not imply a software implementation is essential in order to facilitate efficient configuration. It enables solutions that can be implemented differently in different product variants, sometimes using software and sometimes using discrete electronic components. To manage software, an intermediate object representing the source code was found to be beneficial.

The information model and supporting infrastructure should map to the system architecture in order to be able to support configuration. The platform must be prepared for the configuration strategy designed in the information model.

6.5.3 The FDoc study

In the FDoc study the documentation of functions at Scania were evaluated through the application of an information model. In the CTA and SAINT studies the proposed information models aim at the configuration of product individuals for analysis and production, the FDoc model targets the documentation on a more general level. The work included modeling of concepts and product data regarding the EE (Electrical-electronic) system with a special focus on the concept of function [Larses & El-khoury 2005b]. The details of the model are presented in section 6.6.

The case study was performed in three stages. First an initial pre-study was performed where the current situation at Scania was investigated and a proposal for the modeling of functions was developed. The second stage concerned presentation and discussion to develop an information model and a suggestion for a documentation approach to functions. In the final stage some practical work was performed to try the proposed modeling approach in practice.

In the final stage of the study, three different user functions were used as case study material: *Adaptive cruise control, parking brake warning & cruise control*. Each of the case studies has served different purposes. Adaptive cruise control was the first case, used as a rather advanced function to explore different concepts and ideas and to establish the information model. Different representations and general modeling guidelines were investigated through this case. The second case study, parking brake warning, provided an example where the ideas to improve the current document structure could be applied. In the process, a complete set of documentation of the studied user function was developed based on the new proposed document structure. The work resulted in a template document as well as an illustrative example document. The cruise control was the third case study. This acted as a test of the improvements of the documentation suggested in the second case study.

Improvements of documentation

The analysis in the case focused on a user function specification (UFS) document utilized at Scania. In the analysis of the current approach for function specification, certain areas of improvement were identified. These are presented below, together with some suggestions on how they could be improved.

Documentation format

As the UFS documentation is mainly performed in Microsoft Word and links between documents are managed by hard coded text in the documents, it is very easy to fail to maintain all links when solutions and documentation changes. This can be remedied by maintenance of links in a machine readable format or by replacing Word with a centralized tool where the documentation is stored (for example a database).

Function representation

The current UFS uses message sequence charts (MSCs) as an extension to the word documents to describe and represent of functions. Each MSC represents a scenario that describes a part of the behavior of the function. It is very difficult to capture <u>all</u> possible scenarios of a function in this manner.

One model representing the complete functionality would be useful. The current UFS does not provide a general overview of the function through a model. A collection of MSCs is required to capture the complete functionality. This makes analysis of the system much more difficult as it takes much time to fully grasp a function. The navigation of the documentation becomes less efficient. One way to improve this is to utilize UML 2.0 activity diagrams. However, with an activity diagram the exact behavior of the system is not necessarily defined. The valid paths of behavior execution are not explicitly shown and for this purpose a collection of MSCs can be useful to identify these valid paths and provide the details of some common scenarios.

Traceability and levels of abstraction

To achieve good traceability, a fine grained documentation structure is required. With the current documentation where there is no division of user functions to smaller parts, the assignment of functionality to components is implicit since it is not strictly defined what part of the function is carried out in what component. The UFS targets the communication needs for distributed functions with limited pointers to the software implementation within the ECUs.

Another problem with the current definition of user function is that it does not cover functions with the software completely implemented in one ECU. These functions are not documented in MSCs. This can be improved by introducing a more fine grained function concept that covers all functionality.

Further, without a fine grained documentation structure it is difficult to find dependencies between user functions due to shared components. Resources shared by several functions may cause unfortunate effects. The alteration of a component to remedy a fault in one function may cause another function to fail. This could be the case for a share CAN-message that is altered to fit one user function and then fails to support a different user function.

Separation of implementation and function

As the current User Function Specification document contains information both about the function and the. Due to this, function documentation cannot be reused if the implementation changes. A new hardware or architecture requires that a new function document is registered.

Further, architecting becomes more difficult as assignment of functions to components must be performed in detail for each user function instead of being performed on a higher level, designing functions and architectures separately and then relate the two.

The simple remedy for these problems is to separate documentation of functions and implementation which enables a clearer process responsibility. Such separation is easier if an information model is available that clearly distinguishes and defines the entities describing both function and implementation.

Results

The current function documentation can be improved in the short term by separating the function specification into two documents with more well defined boundaries between functions and the implementation, as prescribed by the information model.

A more fine grained function description model allowing the explicit allocation of functions to hardware has also been introduced. The model utilizes representations of UML 2.0 as a graphical notation for describing functions. A function representation based on activity diagrams, has specifically been proposed.

An information model representing the main concepts and entities envisaged in the product being modeled has been developed. The information model was useful to identify problems with boundaries in the documents and also to identify the need for communication of information regarding specific information entities. In the long term, this information model should become the basis for all documentation. This approach helps in the management of dependencies and relations between the information objects, reducing sources of duplications and inconsistencies in the documentation set. Roles can be associated to information objects, distributing the responsibilities and interactions between the various roles within the organization.

With a proper view of the relations between concepts and product data in the organization, both the quality of information and man-hours invested in information management will improve, resulting in a dependable and cost-efficient lifecycle process.

6.6 An Information Model for Embedded Systems

In the line of work with the different case studies different parts of an information model was developed. The most elaborate model was provided in the FDoc study and this model is accounted for and discussed in this section. Some references to the other case studies and some general possible extensions are also provided. In the FDoc study the purpose of the information model was to serve as a basis for the analysis of documentation of embedded system development with a certain concern for functions.

An information model is a representation of the main concepts and entities envisaged in the product being modeled. Here, these entities are termed information model objects or IMOs (*IMObjects in* Figure 6-18). When using a class diagram to represent the information model, IMOs are modeled as classes, and the relations between the IMOs (such as dependencies, aggregations, etc.) are modeled through the associations between classes. A product is described by instantiating IMOs and their relations.

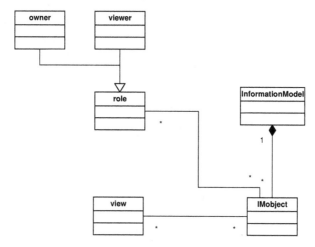

Figure 6-18 Basic concepts

One way to understand and manage the information model is to group IMOs into different *views*. A view hence forms a subset of the information model. Each view is specified to target a specific audience, using that audience's familiar language (viewpoint), and concentrating on that audience's concerns, or particular aspects of the system.

A *role* controls access to IMOs. It is suggested that each IMO is associated with two types of roles, *owner* and *viewer* roles. Owners are allowed to view, create, modify and delete their designated IMOs. Viewers are allowed to view their designated objects. An IMO may exist in several views, but an IMO should have one and only one owner.

While roles are associated to IMOs, these roles indirectly relate to the views grouping IMOs, and certain rules need to be established on how roles access a complete view. To obtain view access to a view, a role should have at least 'viewer' access to ALL the IMOs making up that view. Also, within a view, a role is only allowed to create, modify and delete IMOs to which it possesses owner role.In the information model proposed here, and shown in Figure 6-19, the three main views of the system are the Functional view, Software view and Hardware view, which are further described in section 6.6.1, 6.6.2 and 6.6.3 respectively. A common pattern exists between each of these views. For this reason, the Function view will be described in details, and the analogy should be drawn for the remaining views. The common pattern detected between the information models used for each of these domains is further detailed in section 6.6.6.

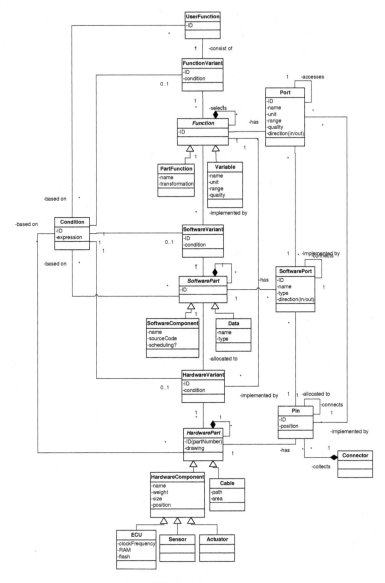

Figure 6-19 The information model

Special attention should be placed on the hierarchical description used within each view. This highlights the fact that there exists no single dominating product structure, and each view describes the system structure/hierarchy from a specific perspective. The User Function view (section 6.6.4) is a special view targeting the product user, and hence focuses on structuring the product functionality from the user perspective. Given the importance of product configurations, each of the above views is further described using a specific variant view. Section 6.6.5 describes the three variant views: FunctionVariant, SoftwareVariant and HardwareVariant views collectively.

It is also important to note that objects do not exclusively belong to one view. For example, the SoftwarePart object belongs to both to the Software view describing the software implementation, as well as the HardwareVariant describing the allocation of software to hardware. Such objects help identify the dependencies that exists between views, calling for special attention for their management, in order to reduce duplication and inconsistencies in the product description. It is important however to define a single view in which the common object can be modified by its owner, while other views are used as viewers of the object.

6.6.1 Function View

The main object in this view is the *Function*. Two sub-types of Function objects are defined: *PartFunction* and *Variable*. A PartFunction object designates certain functionality that given a certain input, produces a certain output. A Variable object designates a transportation link that manages certain data internally and provides access to this data to connected PartFunctions.

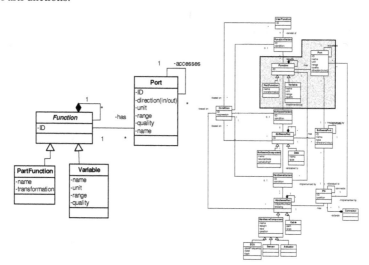

Figure 6-20 Function view

These object types are arguably similar, taking certain input and producing output. The difference lies in the intention of each type, which is ultimately decided upon by the user. A variable object differs from a PartFunction object in that its main purpose is the data transfer it performs, while its functionality (the transformation of the data during the process) becomes a side effect. The PartFunction's main purpose is to transform its input data to produce some output data, where the transformation is not seen as a transfer of data.

A Function can be decomposed into a set of (sub-)Functions, forming a hierarchical product structure, leading to two different descriptions: the *internal definition* (white-box definition) and the *external/interface definition* (black-box definition) [El-khoury 2005].

The interface definition of a Function is an extract of the internal definition, and is defined by a set of ports. A *Port* forms part of the interface of its object and acts as a placeholder for a subset of its object's externally accessible properties. It is through ports that an object interacts with its external environment. In order to externally reveal the internal properties of an object, an object's port establishes an *interface relation* to the port of the internal Function with the properties of interest.

Function representation

There are many different ways to describe the network of functions for a given system. Three such approaches have been evaluated, namely Activity diagrams (AD), Data Flow Diagrams (DFD), and Message Sequence Charts (MSC).

The advantage of using an AD over an MSC is that the former represents all the behavioral scenarios of the functionality using a single diagram. MSCs can only be used to describe a certain sequence of events for a given functionality, but are quite hard to use for a complete definition. It is possible to use a set of MSCs as a complementary specification to an AD, when it is desired to discuss specific scenarios. In this case, the MSC should be checked for consistency with the defining AD, and changes to the AD specification should be automatically reflected in any MSC defined.

The DFD can also be used to fully define a function. A DFD differs from an AD in that focus is placed on the data flow between functions in the network, while the control flow between them showing the sequence of activities is not specified. An AD in which the control flow is excluded can serve this purpose, making the use of a DFD redundant.

Given this reasoning, the AD approach is suggested. However, as the AD can be used so freely and in different ways, rules for how it should be used are needed. These rules should ensure that the required data flow or control flow are properly modeled and complete.

6.6.2 Software View

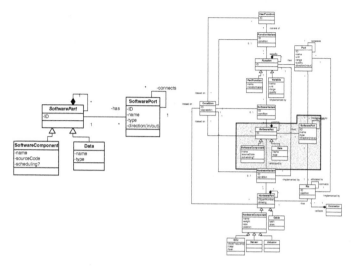

Figure 6-21 Software view

Similar to the Function view, the main object in this view is the *SoftwarePart*, with two sub-types: *SoftwareComponent* and *Data*. A SoftwareComponent object designates a source code module that given a certain input, produces a certain output. A Data object designates a data storage facility that manages certain data internally and provides access to this data to connected SoftwareComponents.

A SoftwarePart can also be decomposed into a set of (sub-)SoftwareParts, forming a hierarchical product structure, leading to two different descriptions: the *internal definition* (white-box definition) and the *external/interface definition* (black-box definition).

The interface definition of a SoftwarePart is defined by a set of SoftwarePorts. A *SoftwarePort* designates a certain internal data item that is externally accessible to other SoftwareParts. The data is described using the name, direction (in, out) and type (int, long, etc.) attributes. Connection relations between SoftwarePorts indicate that the input data of one SoftwarePort is the output data of the other. Since ports of SoftwareComponent objects can only connect to ports of Data objects, a connection relation indicates that the connected port of a SoftwareComponent exchanges its data via the connected Data's SoftwarePort. A SoftwarePort connected to more than one SoftwarePort indicates that the data on that port is transmitted through all of the connected SoftwarePorts. Interface relations indicate that the related port of the internal object is available for external interface.

6.6.3 Hardware View

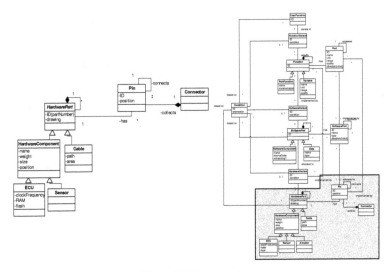

Figure 6-22 Hardware view

Similar to the Function view, the main object in this view is the *HardwarePart*, with two sub-types: *HarwareComponent* and *Cable*. A HardwareComponent object designates a physical block occupying a certain amount of space. It is simply modeled as a 3-D square box and its attributes describe its geometrical dimensions and position. A Cable object designates a single cable with a certain geometrical path. Its attributes describe its spatial path and area.

A HardwarePart can also be decomposed into a set of (sub-)HardwareParts, forming a hierarchical product structure, leading to two different descriptions: the *internal definition* (white-box definition) and the *external/interface definition* (black-box definition).

The interface definition of a HardwarePart is defined by a set of pins. A *Pin* designates a spatial location at which the HardwarePart can be connected to other HardwareParts. Interface relations indicate that the pin of the internal HardwarePart is available for external connections. A *Connector* is used to group related pins into a single entity. A simpler information model might consider a connector to be a special type of HardwarePart.

6.6.4 User Function

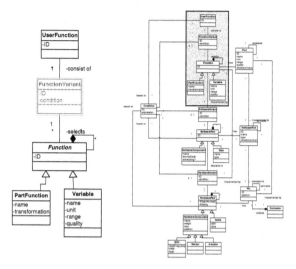

Figure 6-23 User function view

A complete system is described using a network of hierarchically decomposed Functions. However, from the user perspective, certain sets of Functions form a clear and valuable contribution that the user can relate to. Such a set is managed in the information model using the *UserFunction* object.

Ignoring Function variants for the moment, a UserFunction is a grouping of Function objects, forming a fully defined specific functionality (just like the hierarchical composition of functions into PartFunctions).

It is important to note that a Function object is not exclusively selected by a single UserFunction. Certain functionality, such as "speed sensing", provides services that can be shared by many user functions. Such function objects are a good example of the interaction and dependencies that must be managed between user functionalities.

6.6.5 Variants

It is necessary to manage variants of functions, software realizations of functionality and the hardware platform in which the software realizations are allocated. Again, a pattern can be found in representing these three variant needs, and in their relation to other objects in the information model.

Given that different variants are available for a particular user functionality, the UserFunctionVariant is used to represent variations. A UserFunctionVariant is a grouping of Function objects, forming a fully defined specific functionality (just like the hierarchical decomposition of functions into sub-functions). A UserFunction becomes a grouping of UserFunctionVariants that provide similar or competing functionality from which the user can choose one. It is important to note that a Function object does not exclusively belong to a single UserFunctionVariant, since certain functionality can be a common part among the various variants of a given UserFunction.

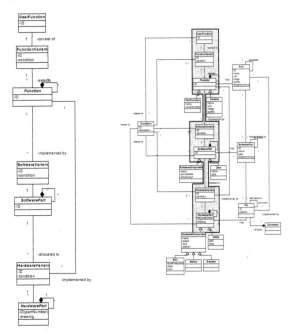

Figure 6-24 Variant view

It may also be the case that different variants exist for how a particular Function is implemented in software. The SoftwareVariant is used to represent such variations, where each SoftwareVariant is a grouping of SoftwarePart objects that equally realize a given Function.

Similarly, different variants can exist for how a SoftwarePart is allocated to hardware or a Function is implemented in hardware. The HardwareVariant is used to represent such variations, where each HardwareVariant is a grouping of HardwarePart objects that equally carry a given SoftwarePart or implement a given Function.

The relationship formed between UserFunction, FunctionVariant and Function is similar to that between Function, SoftwareVariant and SoftwarePart, which in turn is similar to that between SoftwarePart, HardwareVariant and HardwarePart. Each of the relationships maps objects of two views to each other through variants. This is illustrated in the figure below, where an object of view Y is mapped to view X similar to the UserFunction, FunctionVariant and Function objects respectively.

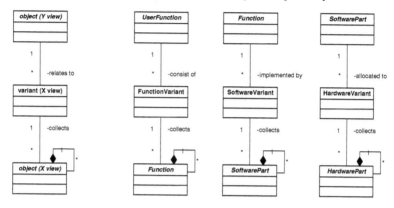

Figure 6-25 Generalized variant

The objects can be interpreted as follows:

- *object (Y view)*: a specific object in one view of the system that is related to other views.

- There is a set of different variants/possibilities to relate this object to the other view (a set of *variant*s).

  o *variant (X view)*: one of the many possibilities to relate the two views.

- For each *variant*, there is a set of collected objects (a set of *X view objects*).

  o *object (X view)*: an object that (partly) realizes the relation for that given variant.

  o *variant* acts as the grouping entity of all the *object*s that together can fully realize the relationship, for that particular *variant* Y.

- An *object (X view)* can be realizing parts of the relationship for more than one *variant* (X view), and more than one *object (Y view)*.

Variant Condition

Enabling logical constraints to be imposed on the possible variants of the system are a desired feature. This is performed using the Condition object. A Variant (FunctionVariant, SoftwareVariant or HardwareVariant) object is related to a Condition object. The Condition object describes the condition for the particular Variant to be valid using the expression attribute, which operates on the related (based on) UserFunction, HardwarePart and SoftwarePart objects.

These conditions could be separately defined for each of the variant types corresponding to three different types of condition objects. In the information model only one type of object is shown collecting the ability to manage function, software and hardware variants.

6.6.6 A generalized domain model

Figure 6-26 describes the common pattern detected in the information models used for each of the domains discussed earlier.

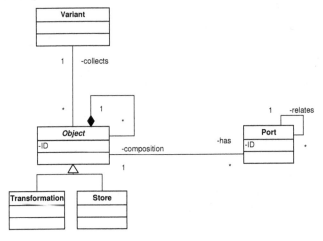

Figure 6-26 Generalized domain model

A hierarchical decomposition of each domain forms a fundamental basis for managing the size and complexity of the system description. A system description is decomposed into a set of objects which are themselves systems that are further decomposed. A hierarchical product structure is thus formed.

This leads to two different descriptions. If viewing the object as the parent object containing other objects, then the *internal definition* (white-box definition) deals with its complete set of properties, which consists of the set of internal objects. This definition defines the object as a stand-alone system and hence needs to be complete irrespective of its surrounding environment. If viewing the object as a composing object of a larger parent object, then the *external/interface definition* (black-box definition) reveals only those properties that need to be shared with the system environment. From the environment perspective, this definition is sufficient to know how the object can be used and related to other objects, while ignoring its internal workings. The completeness of the internal definition and the abstraction of necessary information to the external environment are achieved via the interface definition [El-khoury 2005].

The interface definition of an object is an extract of the internal definition, and is defined by a set of ports. It is through ports that an object interacts with its external environment. In order to externally reveal the internal properties of an object, an object's port is related to an internal object with the properties of interest [El-khoury 2005].

Note that an object interface can only specify how the object can possibly interact with other objects in the system, and not the actual interaction with other objects, since such information is beyond the scope of the object itself. To specify the interaction, objects must be classified as either transformations or stores. Transformations connect stores through interfaces. This ensures the completeness of an object definition, and the independence of object definitions from each other. An object should be fully defined irrespective of the other objects it interacts with. Similar reasoning has been found in several domains of engineering. The need to specify the object interactions as first-class elements (applying transformations and stores) has been identified in software engineering and is further discussed in [Clements et al 2002]. Another example is given by the theory of transmission line elements, initially applied for fluid mechanics [Auslander 1968]. This theory has been translated to electrical network engineering [Johns & O'Brien 1980] and also mechanical engineering [Krus 1999]. The reasoning is also applied in the modeling language Modelica that distinguishes *flow* and *through* variables. In software engineering the pipes and filters design pattern provides another analogy.

In the information model, both transformations and stores are specified as specializations of the object type, allowing the concepts of hierarchical decomposition, variants and interface to be shared by both objects.

6.6.7 Dependability Analysis Extensions

The provided information model also needs to be complemented with objects related to verification and validation, as well as for safety related analysis. These issues have not been studied in the case study but possible extensions are discussed here.

Three additional object types are introduced as possible extensions as shown in Figure 6-27. A *hazard* object related to functions, a set of *verification and validation (V & V)* object related to several objects, and a *failure mode* object related to hardware components. Note that these objects should be adapted to fit the actual test process and safety analysis process. The discussion here is only to provide an example and illustration of how the information model might fit to these aspects.

Risk analysis

Hazards can be identified and related to Functions that may have hazardous behavior. The identification of such hazards will allow a redesign of functions or place requirements on the reliability of implementing components, possibly indicating redundancy.

Hazards can be introduced in the information model as an object related to Functions. The hazard objects must be identified through risk analysis.

Hazard analysis integrated in functional descriptions decomposed in an object oriented fashion have been proposed by Johannessen et al [2001] who tried the approach at Volvo Car Corporation with good results. In a case study they applied analysis on UML models. Automated fault tree analysis integrated in Simulink modeling has been performed by Papadopoulos and Grante [2003] who believe that a more model based approach can improve the safety analysis process. The diagrams used in these cases can easily be mapped to our information model and the functional view.

Failure modes and effects analysis (FMEA)

With a well defined information model it becomes easier to perform FMEA. The implications on function level for a faulty component can easily be evaluated. The FMEA can be related to each component in the implementation and can be seen as a part of the verification efforts in the model, possibly the FMEA can be seen as an attribute to the objects themselves or it can be modeled with an object of its own. A possible extension with a failure mode object is shown in the Figure 6-27.

Automated FMEA generated from fault trees extracted from Matlab/Simulink has been explored by Papadopoulos, Parker and Grante [2004].

Testing

Verification and validation can be performed based on several views and several levels of abstraction. Verification objects can be connected to several of the objects in the information model depending on the verification strategy. The effort invested in different verification methods and the responsibility for approval of tests can be modeled by adding relations to different parts of the information model. A possible set of additions with verification and validation (V & V) objects are shown in Figure 6-27.

6.6.8 Comparing the information models of the cases

The information models have been developed based on experiences collected in each case. The initial model was provided in the CTA case where the separation of function and implementation was examined.

In the CTA case the *functions* and *communication links* are actually derived from the documentation of the software supplied by the MSCs. This means that the concepts of function and software are used interchangeably, and no separate object to represent software is used. This reduction in the model is possible as the case uses documentation of the existing implementation where the assignment of functions to software already has been performed, and the required data is only the communication patterns created by the allocation of software to hardware.

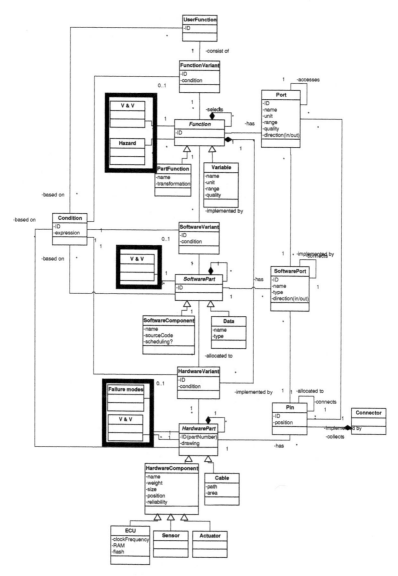

Figure 6-27 The extended information model

In the SAINT case the concepts of function, software (in terms of source code) and hardware are duly separated to provide support for configuration management. The configuration of a product individual, represented by the *customer configuration* object, is based on the selection of a given functional content, similar to the CTA case. This functional content is traced through the source code of the software to the final software and hardware configuration. A mapping immediately to the implementation is also provided for non-functional selections such as performance characteristics of a given component. The software source code object is required to provide configuration support for software components.

With software, hardware and functions separated a more detailed model was developed in the FDoc case. However, the FDoc model lacks the product individual object that is a part of both the CTA and the SAINT models. This object is required to provide support for configuration and analysis of a given vehicle and can easily be added to the FDoc model. In the CTA case, shown in Figure 6-16, a product individual is mapped to a selection of *functions* based on the *scenario* object. The *scenario* object selects a set of *functions* and *communication links* which could be mapped to the *PartFunctions* and *Variables* of the FDoc model shown in Figure 6-19. In the SAINT case the product individual is also immediately related to the implementation objects which also can be included in the FDoc model.

In the SAINT model the separation of component and connector, as exemplified by *software component* and *data* in the FDoc model, is not explicit. This is not required as the model is not utilized for analysis but rather for configuration management where connectors and associations are known and can be implicit in the model.

An important point is that different views and purposes require different object to be included in the information model. However, all the views and models must be integrated in a common information model that provides a basis for the design, integration and configuration of the system.

The architecture design method proposed in chapter 5 can be applied in several ways to align the structures of related views are in a manner that makes the complete system modular. The method requires that all objects that are to be coordinated are recognized in the information model. If concepts and objects are explicitly defined the method can support clustering of functions for implementation in common software and hardware components, as well as clustering of software for allocation to common hardware components.

The information model has developed organically from the needs encountered in the different research studies. Different modeling technologies were studied to provide theoretical support for the derived information model. The architecture analysis provided an initial need for structured data and the information modeling was further developed through the SAINT and FDoc cases. The derived model should be compared to the other, more mature, information modeling efforts such as some application protocols on STEP. Merging the model derived here with such technologies would be an interesting direction of future work.

6.6.9 Conclusion

Providing an information model is useful. The information model should be shared by all views of the system, where each view only utilizes parts of the model. The model clarifies the responsibilities in product development and documentation, and provides traceability for analysis and configuration management.

The experiences from the partial information models developed in the case studies provide an indication of the minimum modeling effort required to support the different purposes. The benefits of separation of function and implementation have been studied in the development of tools inspired by the conceptual Monty model, both for architecture design in the CTA case and also in the development process in the SAINT case. The need for separate representation of software in the shape of source code was identified in the SAINT case. These experiences were combined and extended with more details in the FDoc case, such as including ports. Further extension and alignment with mature technologies would be an interesting line of further work,

The information objects needed in the model vary according to the purpose of the model. Different views require different models but they can all be connected. The case studies utilized the information model for structuring documentation (FDoc), to support system level analysis (CTA), and for configuration management in an invented production and reconfiguration process (SAINT).

The information model must map to model representations of the information objects. These representations can be simple text sections but they can also be a number of graphical models such as UML-diagrams. The representations are collected in defined work products (i.e. documents) in the lifecycle process. A clear mapping from model representations to the information model helps in the management of dependencies and relations between information objects in work products, reducing sources of duplications and inconsistencies in the documentation set.

With a proper approach to product data management, information quality improves and the accessibility of information is enhanced. Further, the dependability can be improved through more accurate analysis of the system. Also, the man-hours invested in collecting and communicating information can be reduced, providing better conditions for building a dependable and cost-efficient vehicle.

7 Results and Conclusion

This chapter provides a summary of the findings in relation to the research questions posed for the Doctoral Thesis. The results are also translated into implications for industry. In the end, some directions for further research are provided.

7.1 Recapitulating the research questions

This thesis has investigated how a cost-efficient and dependable automotive embedded system should be designed and built. Considerations both for product properties, such as the system architecture, as well as more methodological issues have been covered.

It is important to recognize that by choosing design methods the architectural options are reduced, and also that a given architecture must be accompanied by a proper process. This thesis has discussed these relationships and proposed a set of design methods that are expected to be appropriate for the requirements of automotive electronics.

The long term objective has been to find guidelines to make well-informed decisions on alternative solutions and design methods. For the doctoral thesis a set of research questions was posed, renewing and slightly modifying the licentiate thesis questions:

1. What properties of a system improve cost-efficiency and dependability?

2. What methods can be utilized to ensure these properties?

3. How is the product architecture related to methods used in the lifecycle process?

4. How can guided strategic choices be made between different alternatives (of solutions and methods)?

5. How should tool support be utilized?

The findings and results of the research are summarized below, in relation to the research questions.

7.1.1 What properties of a system improve cost-efficiency and dependability?

The second chapter provides a background on automotive electronics in general, the technologies utilized and their applications. In the third chapter the requirements on dependability and cost are thoroughly explored. It is found that cost must be minimized across the product lifecycle and can be both *volume related*, mainly based on the production cost, and *structure related*, concerning development and procurement. Further, *reliability*, *maintainability* and *safety* are distinguished as important dependability attributes for the automotive sector.

Reliability, discussed in section 3.4, is a well known dependability attribute, and so are methods to improve reliability by adding redundancy in some form. A difficulty is however to properly identify the reliability requirements. Some applications require a high level of reliability that cannot be negotiated, such as safety-critical applications, while in others the cost of increasing the reliability can be balanced with the cost of replacing a faulty component. Balancing the reliability in a cost-efficient manner requires proper analysis of both component cost as well as structure related costs.

The development of reliability requirements related to safety-critical applications is immediately linked to the *analyzability* of the system. As discussed in section 3.5, safety is ensured by proper analysis where critical components are identified. The critical components must either be very reliable or an alternative solution must be found. A good safety assurance process requires that system level hazards can be traced to component level faults and failures, and also that the consequences of propagation of faults and failures can be analyzed. Common techniques include failure modes and effects analysis (FMEA), fault tree analysis (FTA) and hazard analysis. Besides the internal analyzability of the system, it is also required that the external environment is properly modeled and understood in order for hazards and the effects of various system behavior to be properly established.

Maintainability, discussed in section 3.6, relates to the possibility to manage the product throughout the lifecycle, both regarding repair and changes. Thereby it is immediately related to structure related costs. Structure related cost-efficiency is expected to improve with a *modular* system architecture. This modular architecture provides clear boundaries in the product and allows easy management of configuration, development, change and replacement of components. Analyzability is also beneficial for structure related costs, improving the ability to perform diagnosis on the system as well as ensuring correct version and variant management.

Volume related cost-efficiency is provided by reducing the cost to produce a product. This can be achieved for an automotive embedded system through a *centralized* system architecture where hardware is reused and a single ECU carries software that implements several functions. However, software centralization may increase the structure related costs if no infrastructure to manage the software is in place. If an infrastructure is lacking, a functional decomposition of the system where one and only one function is mapped to one and only one hardware unit may be beneficial as the structure related costs can be reduced. A balanced centralization allows optimization between volume related and structure related costs, selecting the degree of centralization.

In summary, three important properties of embedded systems that improve dependability and cost-efficiency are modularity, analyzability and balanced centralization.

7.1.2 What methods can be utilized to ensure these properties?

There are two main methods that are proposed in this thesis to support cost-efficiency and dependability by ensuring modularity, analyzability and balanced centralization. The first is *modular architecture design* and the second is *model based development*.

An architecture design method to achieve modular embedded systems is proposed in chapter 5. The method includes considerations for both analysis and synthesis of solutions. The method can be applied for both architecture strategies mentioned in chapter 2, for a centralized architecture design as well as for a functionally decomposed architecture. The beneficial effects of a modular system are achieved regardless of the degree of centralization. The modularization method may also support decisions regarding if a given function should be centralized or allocated to a separate hardware depending on how intensely the function is related to other functions in a cluster, thus supporting the balancing of the centralization of the embedded system.

Model based development, discussed in chapter 6, is seen as a key method to support analyzability. With a proper information model, traceability throughout the product is supported. This traceability provides a basis for configuration management as well as system analysis. Safety analysis is one important aspect where analyzability is required. Other aspects of information model usage include architecture design through keyfigure analysis, and support for more advanced diagnostic systems. Formalized configuration management enables the design of a more centralized architecture and also supports the management of modularity in software. Some general principles for model based system engineering and an information model are proposed in chapter 6.

A well defined modular architecture design supported by a model based development process would provide a very strong basis to cope with the complexity of future automotive embedded systems.

7.1.3 How is the product architecture related to methods used in the lifecycle process?

The two architecture strategies mentioned in chapter 2, *software centralization* and *functional decomposition* are related to the methods utilized in the lifecycle process. The two design methods proposed to enhance cost-efficiency and dependability, *modular architecture design* and *model based development*, have different relations to the architecture strategies.

As already mentioned, modularization can be applied regardless of the chosen strategy and the proposed architecture design method can actually support decisions on whether to centralize or not. The modularization method can provide complete control unit modules as well as software modules, supporting design for both strategies. However, achieving good modules is expected to be more important for a functional decomposition strategy, considering the focus on decomposition. Therefore, the modular architecture design method is most important for this strategy.

Model based development can also be applied regardless of strategy. However the drivers for a model based approach change depending on the strategy. Complex relations are expected to increase the need for model based development. With a centralization strategy the internal structure in the control units become more complex and model based support must be applied at this level managing the software architecture as modular components. A strategy to develop a functionally decomposed architecture is more focused on providing well defined architected modules in the hardware to be developed, manufactured and maintained. With a functional decomposition the complexity is transferred to the cabling and communication patterns in the network. Integration becomes a complex problem and requires a higher degree of model based support.

An architecture approach that focuses on formalization and structuring of software is to implement a middleware based software architecture platform. One such example is the Autosar initiative presented in section 2.4, which aims for a broader, standardized platform solution.

7.1.4 How can guided strategic choices be made between different alternatives (of solutions and methods)?

Some possible alternative strategies have been identified, both regarding architectures and also regarding development methods. The architecture strategy concerns the degree of centralization of the software in the embedded system architecture. The method strategy concers to what extent a model based development process should be utilized.

Guidelines for selecting strategies have been proposed. On one hand by reasoning about the implication of different architectures for the architecture strategy, and also by introducing a set of drivers for model based development to support the choice of method strategy.

Regarding the architecture strategy, volume related costs are reduced by introducing a more centralized architecture as the number of ECUs can be reduced. Structure related costs are reduced by functional decomposition. In general, both trucks and cars are mainly focusing on volume related costs, however there is a difference between them. Cars are produced in higher numbers and are even more susceptible to volume related costs, while structure related costs can be distributed across a higher number of units. Also, trucks have a longer mileage lifespan which requires more focus on the structure related costs regarding service and repairs. This suggests that car manufcturers should employ a software centralization strategy regardless of the incurred structure related costs. While, truck manufacturers should centralize if tools and processes to manage the software over the product lifecycle are in place in the organization, reducing the structure related costs.

For model based development five drivers have been identified, thoroughly treated in chapter 4. Model based development should be used for a product that is *mature* (not innovative), *standardized* (not customized), and complex in terms of *richness*, *heterogeneity*, and *conflicting requirements*.

The strength of these drivers differs between cars and heavy vehicles. Due to the consumer orientation of cars, the innovation rate and richness of systems are greater than for heavy vehicles that are more commercially oriented. Convenience and comfort functions in the premium segments of cars are not commercially viable for heavy vehicles that are seen as production facilities where excessive costs are immediately cut. Further, heavy vehicle manufacturers develop more components in house and in collaboration with suppliers whereas car manufacturers are more reliant on standards and standard components from the suppliers. Based on this, car manufacturers face stronger drivers and it is not surprising that they are leading the way in model based development.

However, with constantly increasing electronic content in the vehicles it is expected that the degree of model based approaches will increase in general. It is then important to know how to apply modeling to meet the right purposes as elaborated in section 4.4.

7.1.5 How should tool support be utilized?

With a model based approach, some tools need to be adopted into the organization and a process incorporating these tools must be established. It is possible to map the purpose of process activities to different services provided by tools, the proper set of services makes tools *effective*. Further it is possible to link these pairs of activities and services to properties of the product that may act as drivers for a model based development process. The properties of the product complicate the given process activities and suggest that tools and models should be used to manage this complexity, or they provide very favorable reuse effects if tools and models are utilized. If the tools can provide services with low utilization of resources they are also *efficient*. An efficient tool delivers results in a quick and resource efficient way.

Tools are based on an infrastructure that must provide proper data handling in order to supply correct services. Proper data handling is the core of a correct MBD approach. Data handling issues refer to the possibility to maintain consistency and traceability as well as supplying a standardized interface to the data supplied by the tool. Consistency of models and model data is required for valid analysis and must be maintained both in time and space.

To select and use proper tools and to integrate them in a correct and efficient manner it is proposed that a meta-model such as an information model is used. The meta-model should be shared across the tools and identify data items that provide equivalent information, as well as data items that are the transformation of a data item in a different tool. Information modeling is discussed in chapter 6. The chosen information model must support all perspectives on the product that can be found in the organization (and beyond). To ensure compatibility of all modeling efforts in the organization they must be approached as a global initiative. The concepts of different parts of the organization should be coordinated through the information model. Two useful information modeling technologies that can represent a meta-level information model are STEP and UML, other supporting technologies can be found in different architecture frameworks.

If the meta-level is properly defined the choice of tool and modeling language can be more freely chosen. However, the tools and languages used must support the selected information model. Often a given tool provides an internal model that restricts how relations can be established across system components. A proper information model supports the alignment of product, process and organization. The product structure can be defined in the information model. The roles and structure in the organization should map to the objects and relations in this information model. The communication patterns in the process should mirror the relations between objects in the information model. If communication patterns are missing they can be patched by model based approaches where integrated tools can ensure that requirements and solutions are properly transferred between people in the organization.

7.1.6 Concluding remarks

As already discussed in the introduction it must be recognized that the contents of this thesis are influenced by the close connection between the author and Scania. The conclusions are partly based on personal experiences within the heavy vehicle industry in general and Scania in particular.

This research project began as an effort to find architectures and improve architecture design methods. After some initial studies the need for a model based development approach was evident. The architecture analysis methods required proper data as input. This data existed in the organization but it was distributed, fragmented, badly formed and occasionally contradicting. Due to this, the need to improve product data management was identified. Usually the data problems encountered by each individual are too small to acknowledge the need for a larger effort, the individual engineer finds it easier and more efficient to build an own database from information collected here and there. In each separate case this is probably true, and this is also a plausible explanation for the nature of the current situation, with data distributed and fragmented in the organization. However, with a global perspective the number of small problems distributed in the organization sums up to higher costs that can be relieved with a better data management approach.

An alternative approach is to reduce the need to rely on good product data management. With a carefully designed modular architecture where relations between product components are simple, less effort is required to describe the requirements on how components interact and the product data problem is turning local. However, the global problem of managing complex data can not be completely avoided.

On the other hand, with a good product data management system that supports versions and variants, the design of the architecture is less restricted by the process and the architecture can be more optimized based on technical decisions.

Addressing the issues of cost-efficiency and dependability is not a task that is performed by posting fragmented improvements in several domains. To really achieve an improvement, strategic decisions are required that ensure that all efforts strive in the same direction. Adopting a model based documentation strategy cannot be achieved by adding a model here and there. The effort must be based on a global initiative in the organization that can coordinate the work and avoid sub-optimization in the process. This thesis tries to point out the broad perspective of how a coordinated effort can be achieved.

7.2 Future work

This thesis opens up a range of possible extensions and opportunities for further research, both in the line of architecture design and analysis, as well as in the field of information modeling.

The proposed architecture design method should be further tested and evaluated in case studies. The proposed metrics for evaluation of modularity can be further examined, and algorithms to ensure optimization of the metrics can be further developed. Other lines of research within the architecture design field would regard the application of the method. Investigations on proper sets of keyfigures and aspects captured in DSMs can be conducted.

An interesting possibility regarding the architecture track would be to further develop the capturing of design rationale through the cluster analysis method and the allocation of weights to the different DSMs. Is there a method to derive aspects (represented by the DSMs) and to find the weights for these aspects, thereby automating the capturing of the design rationale? Further, with a similar set of aspects captured in DSMs, how does the allocation of weights change between companies and industries? How does the design rationally vary?

The information modeling can also be further studied. The basic model can be elaborated and improved. A more elaborate information model that carries the different aspects of embedded systems in automotive applications can be developed. Views considering parts of the information model can be defined and related to roles in a given process. The partial models accounted for in this thesis developed organically from the needs encountered in the different research studies. Integrating and adapting the derived model to mature information modeling frameworks such as STEP, SysML, AADL and EAST is an interesting direction of future work.

With a well defined set of views it is possible to consider what tools that can support the defined views and evaluate their suitability for a given purpose. Also, general tools integrating other tools and managing the core information model can also be evaluated or developed.

Another interesting direction of further research is to examine the alignment of product, process and organization across the entire product lifecycle. The topic has been partly scratched upon in this thesis but it is possible to further explore the strategic aspects of architecture design, and also to further investigate the details of the processes outside the development process.

There are also some general problems that remain to be solved. An efficient modular infrastructure for configuration management of automotive software is not yet established. Achieving a modular architecture supports a process of continuous development. However, a modular architecture is not the complete solution. Providing an efficient and effective process with supporting methods for continuous development, production and maintenance is another challenge.

7.3 Implications for industry

Based on the results of this thesis it is possible to formulate a plan of action for an industrial player in the automotive industry. The importance of a clear strategy with a complete lifecycle perspective to coordinate efforts has been mentioned. There are two main strategies that must be defined, the *architecture strategy* and the *model based development strategy*.

For the architecture strategy, there are two main alternatives. These alternatives are related to the distribution of control and thus the allocation of software. One can either pursue a software centralization strategy, collecting software on a reduced number of ECUs, or a functional decomposition strategy, utilizing dedicated hardware for each function. Centralization lowers the volume related costs in production but may introduce structure related costs. A recommendation would be to look for the possibility to centralize, but be aware of the structure related costs. The amount of software centralization should be selected based on an analysis of the cost impact considering both perspectives (structure and volume). Heavy vehicles are in general more vulnerable to structure related costs while cars are more dependent on the volume related costs.

Regardless of the chosen degree of software centralization it is beneficial to design your architecture with the modular architecture design method proposed in chapter 5. The method can be applied to both functions and software and helps to optimize the software and hardware architecture in relation to functions, software and strategic aspects. In addition, the method helps to make the architecture design rationale explicit. To be able to use the method a basic modeling of functions is required.

Model based development can be applied at different levels of devotion. In general, some basic level of model based development must be established to manage the increasing complexity of automotive embedded systems. At the lowest level some basic models to support analysis can be introduced in order to ensure the dependability of the system. With extended model support an information model should be produced, simple or elaborate depending on the MBD strategy. The degree of MBD usage should be in line with the strength of a set of proposed drivers. A model based approach is encouraged by a complex, mature, and standardized product. Car systems do in general use more electronics and standards providing stronger drivers for a model based approach compared to heavy vehicles.

The maturity and usage of model based development in the automotive industry today are lower than the drivers recommend. This suggests that initiatives to promote model based development should be launched. With a proper approach to product data management, information quality improves and the accessibility of information is enhanced. This results in better engineering through a dependable and cost-efficient lifecycle process, providing excellent vehicles at affordable prices. If this model based approach is complemented with first-class methods for architecture design, success is inevitable.

References

Adamsson N. 2003. *Modellbaserad mekatronikutveckling och kompetensintegration - En komparativ fallstudie inom svensk fordonsindustri*, Technical Report TRITA-MMK 2003:40 ISSN 1400-1179. Royal Institute of Technology, KTH, Stockholm, 2003.

Adamsson N. 2004. Model-based development of mechatronic systems – Reducing the gaps between competencies? *In procedings of TMCE 2004, The Fifth International Symposium on Tools and Methods of Competitive Engineering.* Lausanne, Switzerland, Volume 1, pp. 405-414, April, 2004.

Adamsson N. 2005. *Mechatronics engineering - New requirements on cross-functional integration*, TRITA-MMK 2005:04, Royal Institute of Technology, Stockholm, Licentiate thesisAdamsson N. 2004. Lic Thesis

Albus J.S. & Proctor F.G. 1996 A Reference Model Architecture for Intelligent Hybrid Control Systems. *Proceedings of the International Federation of Automatic Control.* San Francisco, CA, 1996.

Altera., 2005. *Industry trends.* http://www.altera.com/end-markets/auto/industry/aut-industry.html accessed 2005-07-07.

Amberkar S, D'Ambrosio J, Murray B, Wysocki J & Czerny B. 2000. A System-Safety Process for By-wire Automotive Systems. *SAE 2000 World Congress.* Detroit, Michigan. March 6-9, 2000 SAE 2000-01-1056.

Amberkar S, Czerny B, D'Ambrosio J, Demerly J & Murray B. 2001. A Comprehensive Hazard analysis technique for safety-critical automotive systems. *SAE 2001 World Congress.* Detroit, Michigan. March 5-8, 2001 SAE 2001-01-0674.

Amman P., Ding W. & Xu D. (2001). Using a Model Checker to Test Safety Properties. *Proceedings Seventh IEEE International Conference on Engineering of Complex Computer Systems.* 11-13 June 2001. p212-221.

AMI-C. 2004. Nissan Multimedia Test Vehicle Incorporates AMI-C Specifications. *Informer.* Vol. 4. November 2004. p1-2.

Andreasen, M.M. 1992. Designing on a "Designer's Workbench" (DWB). *In Proceedings of the 9th WDK Workshop*, Rigi, Switzerland. 1992

Aragane Y & Tsuji Y. 2002. Development and Evaluation of Information Communication Control Systems for Reducing Driver Distraction. *SAE Convergence 2002.* Transportation Electronics. Detroit, MI. October 21-23. 2002. SAE 2002-21-0044.

Auer G. 2004. Mercedes ditches glitches with electronics. *Automotive news Europe.* 2004-05-31.

Augustine N R. 2000. Today... Tomorrow... of multidisciplinary systems of systems. *IEEE Aerospace and Electronics Systems Magazine.* Vol 15. No 10. October 2000. p 137–144.

Auslander D.M. 1968. Distributed System Simulation with Bilateral Delay-Line Models. *Journal of Basic Engineering.* Transactions of the ASME, June 1968. p195-200.

Automotive Intelligence. 2001. *Mercedes-Benz F400 Carving.* Available at www.autointell.com/nao_companies/daimlerchrysler/mercedes/mercedes-f400/merc-f400-carving-01.htm accessed 2003-01-17.

Autosar. 2005. *AUTOSAR Web Content.* V22.8. Available at http://www.autosar.org/download/AUTOSAR_Web_Content_V22_8_f.pdf accessed 2005-08-26.

Avizienis A. 1997. Toward systematic design of fault-tolerant systems. *Computer.* Vol 30. No 4. April. p51-58.

Axelsson J. (2001). Unified Modeling of Real-Time Control Systems and their Physical Environments Using UML. *Proceedings 8th International Conference on the Engineering of Computer Based Systems.* p18-25, Washington, April 17-20, 2001

Bate R., Kuhn D. & Wells C. et al. 1995. *A Systems Engineering Capability Maturity Model, Version 1.1,* (SECMM-95-01|CMU/SEI-95-MM-003). Pittsburgh, PA:Carnegie Mellon University, Software Engineering Institute, November 1995.

Beck R, Bracklo C, Faulhaber G & Seefried V. 2001. Backbone-Architektur: Vom zentralen Gateway zur systemintegrierenden kommunikationsplattform. *10 Internationaler Kongress, Elektronik im Kraftfahrzeug.* Kongresshaus Baden-Baden. 27-28 September, 2001.

Berkeley. 2004. "The Ptolemy Project.", Internet site available February 2004: http://ptolemy.eecs.berkeley.edu/

von Bertalanffy, L. 1969. *General System Theory.* New York: Brazilier.

Bishop C.M. 1995. *Neural Networks for Pattern Recognition.* ISBN 019-853864-2. Oxford University Press. Oxford.

Blackenfelt M. 2000. "Modularization by Relational Matrices - a Method for the Consideration of Strategic and Functional Aspects", *Proceedings of the 5th WDK Workshop on Product Structuring.* January 2000, Edt. Riithahuhta A., Pulkkinen A., Springer-Verlag

Blackenfelt M. 2001. *Managing complexity by product modularisation.* PhD Thesis. Department of machine design, Royal Institute of Technology (KTH). TRITA-MMK 2001:1, ISSN 1400-1179, ISRN KTH/MMK/R—01/1—SE. Stockholm. Sweden.

Blixt D, Brikho S, Bråkenhielm E, Cedergren U, Cronebäck Ö, Edvinsson L, Eloranta T, Forséll S, Hallberg M, Karlsson N, Olsson A, Rödén M, Steiner A, Wängdahl J, Öhlund D, Öhrvall M. 2005. *Project SAINT,* Technical Report TRITA-MMK 2005:26 ISSN 1400-1179. Royal Institute of Technology, KTH, Stockholm, June 2005. (In Swedish)

BMW. 2005. BMW. *Technology guide.* Available at http://www.bmw.com/ accessed 2005-05-16.

Bock, C. 2003. UML 2 Activity Model Support for Systems Engineering Functional Flow Diagrams. *Journal of the International Council on Systems Engineering.* Vol. 6, No. 4, 2003

Bock T, Feather K Holfelder W & Lundberg J. 2002. Internet-Based Infotronic Systems Technologies and Markets. *SAE Convergence 2002.* Transportation Electronics. Detroit, MI. October 21-23. 2002. SAE 2002-21-0045.

Bracewell R.H. & Wallace K.M. (2003) A Tool for Capturing Design Rationale. *14th International Conference on Engineering Design. ICED 03.* Stockholm, August 19-21, 2003

Briere D & Traverse P. 1993. AIRBUS A320/A330/A340 Electrical Flight Controls: A Family of Fault-Tolerant Systems. *Digest of Papers, The Twenty-Third International Symposium on Fault-Tolerant Computing.* August 1993

Brooks R. A. (1986). A Robust Layered Control System for a Mobile Robot. *IEEE Transactions on Robotics & Automation.* Vol RA-2. No 1. March 1986. p14-23.

Brown A., Chung L. & Patterson D. (2002). Including the Human Factor in Dependability Benchmarks. *International Conference on Dependable Systems and Networks. DSN 2002.* Washington D.C. USA June 23-26, 2002.

Bryan M.G. & Sackett, P.J. 1997. The point of PDM [product data management]. *Manufacturing Engineer,* Vol. 76, No. 4, p161 – 164. Aug. 1997.

Buede, D. 2000. *The Engineering Design of Systems: Models and Methods.* John J. Wiley & Sons.

Burr H., Vielhaber, M., Deubel, T., Weber, C., Haasis, S. 2004. CAx/EDM Integration – Enabler for Methodical Benefits in the Design Process. *Proceedings of Design 2004.* 8th International Design Conference, Dubrovnik, Croatia (2004) p833-840

Butler D. 2002. Launching Advanced Automatic Crash Notification (AACN): A New Generation of Emergence Response. *SAE Convergence 2002.* Transportation Electronics. Detroit, MI. October 21-23. 2002. SAE 2002-21-0066.

Checkland, P. 1999. *Systems thinking, Systems practice: Includes a 30-Year Retrospective.* John Wiley & Sons.

Chen D. (2001). *Architecture for systematic development of mechatronics software systems.* Licenciate thesis. Department of machine design KTH. TRITA-MMK 2001:06, ISSN 1400-1179, ISRN KTH/MMK—01/06—SE. Stockholm. Sweden.

Clements P., Bachman F., Bass L., Garlan D., Ivers J., Little R., Nord R. & Stafford J. 2002 *Documenting software architectures: Views and beyond.* Addison Wesley, Reading, MA, 2002.

CM-SEI. 2004. *Capability Maturity Model® for Software (SW-CMM®):* Carnegie Mellon Software Engineering Institute. http://www.sei.cmu.edu/cmm/

Coelingh E, Chaumette P & Andersson M. 2002. Open-Interface Definitions for Automotive Systems: Application to a Brake-by-Wire system. *SAE 2002 World Congress.* Detroit, MI. March 4-7. 2002. SAE 2002-01-0267.

Collins R, Bechler K & Pires S. 1997. Outsourcing in the automotive industry: From JIT to Modular Consortia. *European Management Journal.* Vol 15. No 5. p498-508.

Connell G. (1989). *A Colony Architecture for a Mobile Robot.* PhD Thesis. MIT Artificial Intelligence Laboratory. 1989.

Cooling J.E. (1991) *Software design for Real-time systems.* Chapman and Hall, 1991. ISBN 1-85032-279-1.

Courtois P.J. & Parnas D.L. (1993). Documentation for Safety Critical Software. *Proceedings 15th IEEE International Conference on Software Engineering.* Baltimore, May 1993, p315-323.

CVS. 2005. CVS - Concurrent Versions System. Open source version control. http://www.nongnu.org/cvs/ accessed 2005-10-13.

Dahlberg E. 2003. Personal Communication. Chairman of the *Road Safety Team* at Scania CV AB.

Daniels, J. & Bahill, T. 2004. The Hybrid Process that combines traditional requirements and use cases. *Journal of the International Council on Systems Engineering.* Vol. 7, No. 4, 2004

D'Avello B & Van Bosch J. 2002. Portable and Embedded Wireless Devices as Conduit for Telematics Applications. *SAE Convergence 2002*. Transportation Electronics. Detroit, MI. October 21-23. 2002. SAE 2002-21-0049.

Davenport T. 1997. *Information Ecology – Mastering The Information And Knowledge Environment*. New York, Oxford University Press.

Davenport T & Prusak L. 1998. *Working Knowledge – How Organizations Manage What They Know*. Boston, Harvard Business School Press.

DBench. 2003. http://www.laas.fr/DBench/ accessed 2003-09-30

DeMarco T.1978. *Structured Analysis and System Specification*. Prentice Hall, Englewood Cliffs, NJ, 1978.

DeVries P, Chrysochoos & Kumthekar S. 2002. The Mobile Wireless Ether – Finding Its Way Into the Automobile. *SAE Convergence 2002*. Transportation Electronics. Detroit, MI. October 21-23. 2002. SAE 2002-21-0046.

Dubrovsky V. 2004. Toward System Principles: General System Theory and the Alternative Approach. *Systems Research and Behavioural Science*. Vol 21. 2004. pp 109-122.

Dutertre B & Stavridou V. A model of noninterference for integrating mixed-criticality software components. *Dependable Computing for Critical Applications*. No 7. Nov 1999. p301-316.

EAST-EEA. 2005. http://www.east-eea.net/ accessed 2005-08-26

EAST-EAA. 2005b. Embedded Electronic Architecture – Definition of language for automotive embedded electronic architecture. Report D3.6, ITEA project 00009. http://www.east-eea.net/ - accessed April 2005.

El-khoury J. 2005. *Towards a Multi-view Modelling Environment for Mechatronics Systems*. Technical Report TRITA-MMK 2005:24 ISSN 1400-1179, ISRN/KTH/MMK/R-05/24-SE. Mechatronics Lab. Royal Institute of Technology. Stockholm.

El-khoury J, Chen D-J & Törngren M. 2003. *A Survey of Modeling Approaches for Embedded Computer Control Systems*. Technical Report, TRITA-MMK 2003:36 ISSN 1400 –1179, ISRN KTH/MMK/R-03/11-SE, 2003.

El-Khoury J & Törngren M. 2001. Towards a Toolset for Architectural Design of Distributed Real-Time Control Systems. *Proc. of Real-Time Systems Symposium (RTSS)*, 3-6 December, 2001, London, pp. 267-276

Emaus B D. 2000. Current Vehicle Network Architecture Trends – 2000. SAE *2000 World Congress*. Detroit, MI. March 6-9. 2000. SAE 2000-01-0146.

Ender M. 2002. Technology and Business Implications for Distributed Telematic Applications. *SAE Convergence 2002*. Transportation Electronics. Detroit, MI. October 21-23. 2002. SAE 2002-21-0047.

Eppinger S., & Salminen V. 2001. Patterns of Product Development Interactions *Proc. Int. Conference On Engineering Design, ICED01* Glasgow, August 21-23, 2001, pp 283-290.

Erixon G. 1998. *Modular Function Deployment – a method for product modularisation*. PhD Thesis, Royal Institute of Technology (KTH), Stockholm, 1998.

ESTEREL. 2003. *Esterel Technologies – Customers*. Press releases. http://www.esterel-technologies.com accessed 2003-10-14

Fantechi A., Gnesi S. & Semini L. (1999) Formal Description and Validation for an Integrity Policy Supporting Multiple Levels of Criticality. *7th IFIP International Conference on Dependable Computing for Critical Applications*. San Jose, Ca, USA. IEEE Computer Society Press, 1999.

Feick S, Pandit M, Zimmer M & Uhler R. 2000. Steer-by-Wire as a Mechatronic Implementation. *SAE 2000 World Congress*. Detroit, MI. March 6-9. 2000. SAE 2000-01-0823.

Fettweis A. 1971. Digital filter structures Related to Classical Filter Networks. *Arch Elek Übertragungst*. Vol 25. No 2. p79-89

Fischer J, Holz E & Møller-Pedersen B. 2000. Structural and Behavioral Decomposition in Object Oriented Models. ISORC-2000: *The 3rd IEEE International Symposium on Object-oriented Real-time distributed Computing*. California, March 2000, p. 368-375.

Flood R.L. 2000. A Brief Review of Peter B. Checkland's Contribution to Systemic Thinking. *Systemic Practice and Action Research*. Vol. 13, No. 6, 2000, p723-731.

FMV. (2001). *Försvarsmaktens handbok för programvara i säkerhetskritiska tillämpningar*. H ProgSäk. M7762-000531. Försvarets Materielverk. Försvarsmakten. Sweden.

Frank R. 2002. Towards the Intelligent Power Network. *SAE Convergence 2002*. Transportation Electronics. Detroit, MI. October 21-23. 2002. SAE 2002-21-0060.

Fredriksson U. (1993). *JAS 39 Gripen crash in Stockholm 1993 Aug 08 report summary*. http://www.canit.se/~griffon/aviation/text/gripcras.htm accessed 2003-07-30.

Freeman T. 2004. Plastic optical fibre tackles automotive requirements. *Fibre Systems Europe in association with Lightwawe Europe*. May 2004. p14

Freund, U., Gurreri, O., Lönn, H., Eden, J., Migge, J., Reiser, M.-O., Wierczoch, T. & Weber. M. 2004. An Architecture Description Language supporting automotive software product lines. *In: The Third Software Product Line Conference, SPLC2004, workshop on Solutions for Automotive Software Architectures*, Boston, Massachusetts, September, (2004)

Fuchs M, Schmerer R & Zeller A. 2002. Comfort and Convenience Features in Luxury Cars. *SAE Convergence 2002*. Transportation Electronics. Detroit, MI. October 21-23. 2002. SAE 2002-21-0052.

Flexray. 2005. *Flexray – The communication system for advanced automotive control applications*. Available at http://www.flexray.com accessed 2005-08-15.

Garlan D, Allen R & Ockerbloom J. 1995. Architectural Mismatch: Why Reuse is so Hard. *IEEE Software*. Vol 12. No 6. November 1995. p17-26.

George R & Wang J. 2002. Vehicle E/E System Integrity From Concept to Customer. *SAE Convergence 2002*. Transportation Electronics. Detroit, MI. October 21-23. 2002. SAE 2002-21-0018.

Gero J.S. 1990. Design Prototypes: A Knowledge Representation Scheme for Design. *AI Magazine*, vol. 11, no. 4, 1990, pp. 26–36.

GM. 2005. *GM Sequel: Reinvented Automobile No Longer Just a Dream*. Webpage http://www.gm.com/company/gmability/adv_tech/100_news/sequel_011005.html accessed 2005-05-16.

Golafshani N. 2003. Understanding Reliability and Validity in Qualitative Research. *The Qualitative Report*. Vol. 8, No. 4, December 2003. p597-607

Guberman S. 2002. Reflections on Ludwig von Bertalanfy's "General System Theory: Foundations, Development, Applications" *Proceedings of the 5th European Systems Science Congress*, Crete, October 2002

Gunzert M & Nägele A. 1999. Component-based development and verification of safety critical software for a brake-by-wire system with synchronous software components. *Proceedings International Symposium on Software Engineering for Parallel and Distributed Systems*. 1999. p134 -145.

Hansen, M., Nohria, N., & Tierney, T., What's Your Strategy for Managing Knowledge? *Harvard Business Review*. Vol.77, No.2. ,1999, pp 106-116.

Harper C. & Winfield A. (1994). A behavior-based approach to the design of safety-critical systems. *IEEE Colloquium on Knowledge-Based Systems for Safety Critical Applications*. 1994. p1-10.

Harter W, Pfeiffer W, Dominke P, Ruck G & Blessing P. 2000. Future Electrical Steering Systems: Realizations with Safety Requirements. *SAE 2000 World Congress*. Detroit, MI. March 6-9. 2000. SAE 2000-01-0822.

Hedlund, G., A model of knowledge management and the N-form corporation. *Strategic Management Journal*. Vol.15, 1994, pp 73-90.

Heinecke H., Schnelle K-P., Fennel H., Bortolazzi J., Lundh L., Leflour J., Maté J-L., Nishikawa K., Scharnhorst T. 2004. AUTomotive Open System ARchitecture - An Industry-Wide Initiative to Manage the Complexity of Emerging Automotive E/E-Architectures *SAE Convergence 2004*. Vehicle Electronics to Digital Mobility. Detroit, MI. October 18-20. 2004. SAE 2004-21-0042.

Hofmann P & Thurner T. 2001. Neue Elektrik/Elektronik Architekturansätze. *10 Internationaler Kongress, Elektronik im Kraftfahrzeug*. Kongresshaus Baden-Baden. 27-28 September, 2001.

Honeywell. 2004. "Dome Home." Internet site available February 2004: http://www.htc.honeywell.com/dome

Hu, Y-S. 1995. The international transferability of the firm's advantages, *California Management Review*, Vol. 37, No 4 Summer, 73-87.

Huber U. & Näher U. 2004. *Failure-free electronics – seven levers for optimized electronics R&D*. McKinsey & Company, Automotive & Assembly. 2004.

Hubka, V. & Eder, E. 1988. *Theory of Technical Systems*. Springer-Verlag. 1988.

Hutton R. 2005. *Mercedes calls back 1.3m cars*. London Times. 2005-04-03

Hydrogen & Fuel Cell letter. 2002. GM Releases First Data, Pictures of Fuel Cell Hy-wire Concept Car, Debut at Paris Auto Show. September issue 2002. www.hfcletter.com/letter/September02/features.html accessed 2003-01-17. ISSN 1080-8019.

Hölttä K, Tang V. & Seering W.P. 2003 "Modularizing product architectures using dendrograms", *Proceedings 14th International Conference on Engineering Design*. ICED 03. Stockholm, August 19-21, 2003

IEEE. 2000. IEEE Recommended Practice for Architectural Descriptions of Software-Intensive Systems. *IEEE Standard 1471-2000*. Approved 21 September 2000. ISBN 0-7381-2518-0.

IDBForum. 2005. *IDB Forum Homepage*. Available at http://www.idbforum.org/ accessed 2005-08-22.

ISOSPICE. 2005. *ISOSPICE – ISO15504 and the SPICE project*. http://www.isospice.com/ accessed 2005-10-14.

Jackson M. 1983. *System Development*. Prentice-Hall, Englewood Cliffs, NJ. 1983

Johannessen P, Grante C, Alminger A, Eklund U & Torin J. 2001. Hazard Analysis in Object Oriented Design of Dependable Systems. *Proc. International Conferece on Dependable Systems and Networks*. IEEE CS Press, 2001, p507-512

Johns P.B. & O'Brien M. 1980. Use of the transmission line modeling (t.l.m) method to solve nonlinear lumped networks. *The Radio Electron and Engineer*. Vol. 50 p59-70 Jan/Feb.

Kahn K. 1996. A definition of interdepartmental integration with implications for product development performance. *Journal of product Innovation management*, March 1996, vol. 13, p137-151.

Kanayama K, Fujiwara J & Yamato K. 2002. Mobile Gateway for Vehicle M2M (Machine to Machine) Connectivity. *SAE Convergence 2002*. Transportation Electronics. Detroit, MI. October 21-23. 2002. SAE 2002-21-0064.

Kant I. 1790. *Kritik der Urteilskraft*. (Critique of Judgment) 1790.

Kanoun K., Madeira H. & Arlat J. 2002. A Framework for Dependability Benchmarking. *International Conference on Dependable Systems and Networks. DSN 2002*. Washington D.C. USA June 23-26, 2002.

Kaplan R S. & Norton D P. 1996. Using the Balanced Scorecard as a Strategic Management System. *Harvard Business Review*, Jan-Feb 1996, p75.

Kaplan R S. & Norton D P. 1997. Why does business need a balanced scorecard? *Journal of Cost Management*, May/June 1997, p5-10.

Karsai G., Sztipanovits J., Ledeczi A. & Bapty, T. 2003. Model-integrated development of embedded software. *Proceedings of the IEEE*. Vol. 91 No. 1 p145 – 164, Jan.2003

Karypis G. 2003. *METIS: Family of Multilevel partitioning algorithm*. http://www-users.cs.umn.edu/~karypis/metis/ accessed 2003-01-23.

Kassakian J. 2002. The 42V PowerNet Story – Challenges of an International Collaboration. *SAE Convergence 2002*. Transportation Electronics. Detroit, MI. October 21-23. 2002. SAE 2002-21-0005.

Kazman R, Klein M, Barbacci M, Longstaff T, Lipson H & Carriere J. 1998. "The Architecture tradeoff analysis method.", *Proceedings of the Fourth IEEE International Conference on Engineering of Complex Computer Systems (ICECCS)*, Monterey, CA, August 1998, p68-78.

Keutzer K. Newton A.R. Rabaey J.M. Sangiovanni-Vincentelli A. 2000. System-level design: orthogonalization of concerns and platform-based design. *IEEE Transactions on Computer-Aided Design of Integrated Circuits and Systems*. Vol 19. No 12. Dec. 2000. p1523-1543.

Knippel E. & Schulz A. 2004. Lessons Learned from Implementing Configuration Management within Electrical/Electronic Development of an Automotive OEM. *Proc. 14th International Symposium of the International Council on Systems Engineering*, Toulouse, June 20-24, 2004.

Koenig, M., Don't fall for that false dichotomy! Codification vs. personalization. *KMWorld*. Vol.10., No. 8., September 2001.

Koopman P. 2002. Critical Embedded Automotive Networks. *IEEE Micro*. Vol 22. No 4. July-August 2002. p 14–18.

Koopman P. 2002b. What's Wrong With Fault Injection as a Benchmarking Tool. *International Conference on Dependable Systems and Networks. DSN 2002.* Washington D.C. USA June 23-26, 2002.

Kopetz H & Grünsteidl G. 1993. TTP – a Time-Triggered protocol for fault-tolerant real-time systems. *23rd IEEE International Symposium on Fault-Tolerant Computing.* FTCS-23, 1993.

Kruchten P. 1995. The 4+1 View Model of architecture. *IEEE Software*, Vol. 12, No. 6, p42-50. ISSN: 0740-7459. 1995.

Kruchten, P. 2005. Casting Software Design in the Function-Behavior-Structure Framework. IEEE Software. No. 2. 2005 p52-58

Krus P. 1999. Modeling of Mechanical Systems Using Rigid Bodies and Transmission Line Joints. *Transactions of the ACME.* Vol. 121. December 1999. p606-611

Laitenberger O, Bell T & Schwinn T. (2002). An Industrial Case study to examine a non-traditional Inspection Implementation for Requirements Specifications. *Proceedings of the Eighth IEEE Symposium on Software Metrics.* METRICS 02.

Lala J. & Harper R. (1994). *Architectural Principles for Safety-critical Real-time Applications.* Proceedings of the IEEE. Vol. 82, No. 1, p25-40, Jan. 1994.

Lambin J-J. (1996). *Strategic Marketing Management.* McGraw-Hill. 1996. ISBN 0-07-709227-9

Laprie J C. 1992. *Dependability: Basic Concepts and Terminology.* Springer Verlag, 1992.

Larses O. 2003a. *Modern Automotive Electronics from an OEM perspective.* Technical Report TRITA-MMK 2003:09 ISSN 1400-1179, ISRN/KTH/MMK/R-03/09-SE. Mechatronics Lab. Royal Institute of Technology. Stockholm.

Larses O. 2003b. *Modern Automotive Electronics from a Dependable systems perspective.* Technical Report TRITA-MMK 2003:38 ISSN 1400-1179, ISRN/KTH/MMK/R-03/38-SE. Mechatronics Lab. Royal Institute of Technology. Stockholm. Sweden.

Larses O. 2003c. Dependable Architectures for Automotive Electronics – Philosophy, Theory and Practice. Licentiate Thesis TRITA-MMK 2003:39 ISSN 1400-1179, ISRN/KTH/MMK/R-03/39-SE. Department of Machine Design. Royal Institute of Technology. Stockholm.

Larses O. 2005a. Applying quantitative methods for architecture design of embedded automotive systems. *Proc. INCOSE International Symposium 2005.* Rochester, NY. July 10-15. 2005.

Larses O. 2005b. *Factors influencing dependable modular architectures for automotive applications.* Technical Report TRITA-MMK 2005:09 ISSN 1400-1179. Royal Institute of Technology, KTH, Stockholm, 2005.

Larses O. & Adamsson N. 2004. Drivers for model based development of mechatronic systems. *In: Proceedings of Design 2004*, 8th International Design Conference, Dubrovnik, Croatia (2004) p865-870

Larses O. & Blackenfelt M. 2003. "Relational reasoning supported by quantitative methods for product modularization", *Proceedings 14th International Conference on Engineering Design.* ICED 03. Stockholm, August 19-21, 2003

Larses O. & Chen D. J. 2003. *The Monty Model for Engineering of Mechatronic Systems.* Technical Report TRITA-MMK 2003:11 ISSN 1400-1179, ISRN/KTH/MMK/R-03/11-SE. Mechatronics Lab. Royal Institute of Technology. Stockholm.

Larses O & El-khoury J. 2005b. *Function Modelling to Improve Software Documentation.* Technical Report TRITA-MMK 2005:25 ISSN 1400-1179. Royal Institute of Technology, KTH, Stockholm, 2005.

Larses O. & El-khoury J. 2005a. *Views on General System Theory.* Technical Report TRITA-MMK 2005:10 ISSN 1400-1179. Royal Institute of Technology, KTH, Stockholm, 2005.

Larses O. & El-khoury J. 2004. Multidisciplinary Modeling and Tool Support for EE Architecture Design. *Proceedings FISITA 2004 30th World Automotive Congress*, Barcelona, Spain, 23-27 May, 2004

Leveson N. 1995. *Safeware: System safety and computers.* Addison-Wesley Publishing Company. 1995.

Leveson N. & Palmer E. 1997. Designing Automation to Reduce Operator Errors. *In proceedings of IEEE Systems, Man and Cybernetics conference.* October 1997.

Leveson N. 2000. Intent Specifications: An Approach to Building Human-Centered Specifications. *IEEE Transactions on Software Engineering.* Vol. 26. No. 1. January 2000. p15-35.

Leveson N. 2002. An Approach to Designing Safe Embedded Software. Proceedings of the Second International Conference on Embedded Software. EMSOFT 2002, Grenoble, France, October 7-9, 2002. *Lecture Notes in Computer Science 2491.* Springer Verlag. p15-29.

Lhamon R. 2002. Migration of Electronics within Integrated Interiors. *SAE Convergence 2002.* Transportation Electronics. Detroit, MI. October 21-23. 2002. SAE 2002-21-0004.

Lions J. 1996. *ARIANE 5 – Flight 501 Failure.* ESA press release, 1996.

Ljung J., Nilsson P. & Olsson U. (ed.). 1994. *Företag och Marknad – Samarbete och Konkurrens.* Studentlitteratur. Lund. 1994.

Lohmar W. 2004. The Virtual Development Process - a Reality at SEAT. *FISITA 2004 30th World Automotive Congress.* Reference F2004F450, Barcelona, Spain, 23-27 May 2004.

Long, J. 2002. *Relationships between common graphical representations in system engineering.* ViTech Corporation, white paper. 2002.

Loose D, Churchill B & Ruthven J. 2002. Business and Technical Aspects of Open and Proprietary Architectures. *SAE Convergence 2002.* Transportation Electronics. Detroit, MI. October 21-23. 2002. SAE 2002-21-0076.

Loureiro G, Leany P.G. & Hodgson M. 2004. A Systems Engineering Framework for Integrated Automotive Development. *Systems Engineering.* Vol. 7, no. 2, p. 153-166.

Lupini C. 2001. Multiplex Bus Progression. *SAE 2001 World Congress.* Detroit, MI. March 5-8. 2001. SAE 2001-01-0060.

Lupini C. 2003. Multiplex Bus Progression 2003. *SAE 2003 World Congress.* Detroit, MI. March 3-6. 2003. SAE 2003-01-0111.

Lygner M. 2002. *Model-based development tool chain at Volvo Cars.* dSPACE News, 1/2002, www.dspace.de.

Mackall D. 1988. *Development and flight test experiences with a flight-crucial digital control system.* NASA Technical Paper 2857, NASA Ames Research Center, Dryden Flight Research Facility, Edwards, CA, 1988.

Malhotra A. 2002. Enabling Technology's Promise – Collaborative Development of Common Requirements for Mobile Information and Entertainment Systems. *SAE Convergence 2002*. Transportation Electronics. Detroit, MI. October 21-23. 2002. SAE 2002-21-0001.

MatrixOne. 2005. *Matrix10: the flexible plm environment*. Webpage available at: http://www.matrixone.com/matrixonesolutions/index.html accessed 2005-10-12.

McElroy J & Goldstein B. 2002. Improving Supply Chain Communications (Dismantling the Tower of Babel). *SAE Convergence 2002*. Transportation Electronics. Detroit, MI. October 21-23. 2002. SAE 2002-21-0007.

McMenamin S. & Palmer J. 1984. *Essential Systems Analysis*. Yourdon Press/Prentice-Hall. Englewood Cliffs, NJ. 1984

Medvidovic N. & Taylor Richard N. 2000. A classification and comparison framework for software architecture description languages. *IEEE transactions on software engineering*. Vol 26. No 1. January. p70-93.

Mercedes. 2003. *Innovation – Research & Technology*. www.mercedes.com/e/innovation/rd/ accessed 2003-01-20.

Millstein S. 2002. vRM (vehicle Relationship Management). *SAE Convergence 2002*. Transportation Electronics. Detroit, MI. October 21-23. 2002. SAE 2002-21-0063.

MISRA. 2001a. *Development Guidelines for Vehicle Based Software*. November 1994. PDF version 1.1. January 2001. www.misra.org.uk/ Downloaded 2002-12-03.

MISRA. 2001b. *Report 2 – Integrity*. February 1995. PDF version 1.0. January 2001. www.misra.org.uk/ Downloaded 2002-12-03.

Modugno F. Leveson N.G. Reese J.D. Partridffe K. & Sandys S. 1996. Creating and analyzing requirement specifications of joint human-computer controllers for safety-critical systems. *Proceedings Human Interaction with Complex Systems HICS '96*. p46-53.

Monacelli G., Sessa, F. & Milite A. 2004. An Integrated Approach to Evaluate Engineering Simulations and Ergonomics Aspects of a New Vehicle in a Virtual Environment: Physical and Virtual Correlation Methods. *FISITA 2004 30th World Automotive Congress*. Reference F2004F406, Barcelona, Spain, 23-27 May 2004.

Montesquieu C L. 1748. *De l'esprit des loix*. (The Spirit of the laws) Genève. 1748.

Mostcooperation. 2005. *Welcome to the MOST cooperation*. Available at http://www.mostcooperation.com accessed 2005-08-15.

Nambisan, S., & Wilemon, D. 2000. Software development and new product development: potentials for cross-domain knowledge sharing. *IEEE Transactions on Engineering Management*. Vol.47, No.2., 2000, pp211-220.

Nasa. 2004: http://step.jpl.nasa.gov/AP233/AP233-overview.html accessed July 2004.

Neumann P. 1993. Saab JAS 39 Gripen Crash. *ACM SIGSOFT Software Engineering Notes*. Vol 18. No 4. October. p. 11.

Newcomb P.J., Bras B. and Rosen D.W. 1996. Implications of modularity on product design for the lifecycle. *ASME DETC, DETC96/DTM-1516*, Irvine, August 18-22, 1996.

Nihtilä, J., 1999. R&D-Production integration in the early phases of new product development projects. *Journal of Engineering and Technology Management*. Vol.16, No.1., March 1999, pp 55-81.

Nonaka, I. 1994. A dynamic theory of organizational knowledge creation, *Organization Science*, Vol. 5, No 1 February, 14-87.

Nonaka, I. & Takeuchi, H. 1995. *The knowledge-creating company*. Oxford University Press, New York, 1995.

Nossal R. & Lang R. 2002. Model-Based System Development. *IEEE Micro*. Vol 22. No 4. July-August 2002. p 56–63.

Näher U. & Radtke P. 2005. Automotive Electronics. McKinsey Company, *Automotive & Assembly*. 2005.

Ogawa T & Morozumi H. 2002. Diagnostic Trends for Automotive Electronic Systems. *SAE Convergence 2002*. Transportation Electronics. Detroit, MI. October 21-23. 2002. SAE 2002-21-0021.

Olson W. 2001. *Identifying and Mitigating the Risks of Cockpit Automation*. Air Command and Staff College Wright Flyer Paper no. 14. Maxwell Air Force Base. Alabama. June 2001.

OMG. 2002. OMG - Meta Object Facility, v1.4. April 2002.

OMG. 2003a. *OMG Unified Modeling Language Specification*. OMG Unified Modeling Language Revision Task Force. Version 1.5. March 2003. Available at www.omg.org.

OMG. 2003b. *MDA Guide Version 1.0*. Object Management Group. Version 1.0. May 2003. Available at www.omg.org accessed 2003-05-21.

OMG. 2003c. *Meta Object Facility (MOF) 2.0 Core Specification. 03-10-04*. Object Management Group. Version 1.0. May 2003. Available at www.omg.org.

Pahl G. and Beitz W., *Engineering Design – a systematic approach*, Springer-Verlag, 1996.

Palmer E. 1995. 'Oops, it didn't arm.' A case study of two automation surprises. *In proceedings of Eight International Symposium on Aviation Psychology*. April 1995. p227-232.

Papadopoulos Y. & Grante C. 2003. Techniques and tools for automated safety analysis & decision support for redundancy allocation automotive systems. *Proceedings 27th Annual International Computer Software and Applications Conference, 2003*. COMPSAC 2003. 3-6 Nov. 2003. p105 - 110

Papadopoulos Y., Parker D. & Grante, C. 2004. Automating the failure modes and effects analysis of safety critical systems. *Proceedings Eighth IEEE International Symposium on High Assurance Systems Engineering*, 2004. p310 - 311

Parnas D.L. & Madey J. 1995. Functional documents for computer systems. *Science of Computer Programming*. Vol 25. p41-61.

Paulk M, Curtis B, Chrissis M. B. & Weber C. 1993. *Capability Maturity Model for Software*, Version 1.1, (CMU/SEI-93-TR-024). Pittsburgh, PA:Carnegie Mellon University, Software Engineering Institute, February 1993.

Peak R.S., Lubell J, Srinivasan V & Waterbury S.C. 2004. STEP, XML, and UML: Complementary Technologies. *Journal of Computing & Information Science in Engineering*. Vol. 4, No. 4, Special Issue on Product Lifecycle Management (PLM), December 2004. p379-390.

Perry D.E. & Wolf A.L. 1992. Foundations for the Study of Software Architecture. *ACM SIGSOFT Software Engineering* Notes. Vol 17. No 4. October 1992. p40-52.

Peterson L. & Davie B. 2000. *Computer Networks a Systems Approach 2nd ed*. Morgan Kaufmann Publishers. San Fransisco, CA, USA.

Pimmler T.U. and Eppinger S.D. 1994. Integration analysis of product decompositions. *ASME Design Theory and Methodology conference*. 1994, DE-Vol. 68, 1994.

Pompei F J, Sharon T, Buckley S & Kemp J. 2002. An Automobile-Integrated System for Assessing and Reacting to Driver Cognitive Load. *SAE Convergence 2002*. Transportation Electronics. Detroit, MI. October 21-23. 2002. SAE 2002-21-0061.

Powell D, Arlat J, Beus-Dukic L, Bondavalli A, Coppola P, Fantechi A, Jenn E., Rabéjac C. & Wellings A. 1999. GUARDS: A Generic Upgradable Architecture for Real-time Dependable Systems. *IEEE Transactions on Parallel and Distributed Systems*. Vol 10. No 6. June 1999.

Prasad B., Wang F. & Deng J. 1997. Towards a Computer-Supported Cooperative Environment for Concurrent Engineering. *Concurrent Engineering*, Vol 5. No 3. Sept. 1997. p233-252.

Quigley C, Tan F, Tang K & McLaughlin R. 2001. An Investigation into the Future of Automotive In-Vehicle Control Networking Technology. *SAE 2001 World Congress*. Detroit, MI. March 5-8. 2001. SAE 2001-01-0071.

Ranville S. 2004. Case Study of Commercially Available Tools that Apply Formal Methods to a Matlab/Simulink/Stateflow Model. *SAE World Congress 2004*. Detroit, MI. March 8-11. 2004. SAE 2004-01-1765.

Rasmussen J. 1993. Diagnostic reasoning in action. *IEEE Transactions on Systems, Man and Cybernetics*. Vol 23. No 4. July/August 1993. p981-992.

Rechtin E. & Maier M. 1997. *The Art of Systems Architecting*. CRC Press. Boca Raton, Florida US. 1997. ISBN 0-8439-7836-2.

Redell O., Elkhoury J. & Törngren M. 2004. The AIDA tool-set for design and implementation analysis of distributed real-time control systems. *Microprocessors and Microsystems*, Vol 28. No. 4, May 2004, p163-182.

Reilhac P & Bavoux B. 2002. Vehicle E/E Architecture: A New Paradigm for Collaborative Product Creation? A Case Study. *SAE Convergence 2002*. Transportation Electronics. Detroit, MI. October 21-23. 2002. SAE 2002-21-0006.

Reichart G & Haneberg M. 2004. Key Drivers for a Future System Architecture in Vehicles. *SAE Convergence 2004*. Vehicle Electronics to Digital Mobility. Detroit, MI. October 18-20. 2004. SAE 2004-21-0025

Renault. 2002. *The Ellypse concept car: "a bubble of optimism in the automotive world"*. www.media.renault.com/data/doc/mediarenaultcom/en/4004_CPZ13-GB.pdf accessed 2003-01-24.

Rieth P & Eberz T. 2002. Reduction of Stopping Distance Through Chassis System Networking. *SAE Convergence 2002*. Transportation Electronics. Detroit, MI. October 21-23. 2002. SAE 2002-21-0027.

Rushby J. 2001. Bus architectures for safety-critical embedded systems. *Springer-Verlag lecture notes in computer science*. Vol 2211. p306-323. Springer-Verlag.

Rushby J. 2002. Using model checking to help discover mode confusions and other automation surprises. *Reliability Engineering and System Safety*. Vol 75. p167-177.

SAE. 2004. *Architecture Analysis & Design Language (AADL)*. SAE Aerospace standard, SAE AS-5506. Issued November 2004. SAE International.

Sage A.P. & Lynch C.L. 1998. Systems Integration and Architecting: An Overview of Principles, Practices and Perspectives. *Systems Engineering*. Vol 1. No 3. p176-227.

Sallee D & Bannatyne R. 2001. Advanced electronic chassis control systems. *SAE Future Transportation Technology Conference.* Costa Mesa, California. August 20-22. 2001. SAE 2001-01-2534.

Saltzer J H, Reed D P & Clark D D. 1984. End-to-end Arguments in System Design. *ACM Transactions on Computer Systems.* Vol 2. No 4. November 1984. p277-288.

Sanfridsson M, Claesson V & Gäfvert M. 2000. *Investigation and requirements of a computer control system in a heavy-duty truck.* Technical Report TRITA-MMK 2000:5 ISSN 1400-1179, ISRN/MMK-00/5-SE. Mechatronics Lab. Royal Institute of Technology. Stockholm.

Schiele P & Durach S. 2002. New approaches in the software development for automotive systems. *International Journal of Vehicle Design.* Vol 28. No 1-3. p241-257.

Schindel, W.D. 2005. Requirements Statements Are Transfer Functions: An Insight from Model-Based Systems Engineering. *Proc. INCOSE International Symposium 2005.* Rochester, NY. July 10-15. 2005.

Schumacher R, Lind R, Ten H & Welk D. 2002. MultiMedia Entertainment: Vehicle Technology and Service Business Trends. *SAE Convergence 2002.* Transportation Electronics. Detroit, MI. October 21-23. 2002. SAE 2002-21-0062.

Schwartz G. 2002. Reliability and Survivability in the Reduced Ship's Crew by Virtual Presence System. *International Conference on Dependable Systems and Networks. DSN 2002.* Washington D.C. USA June 23-26, 2002.

Scobie J, Maiolani M & Jordan M. 2000. A Cost Efficient Fault Tolerant Brake by wire Architecture. *SAE 2000 World Congress.* Detroit, MI. March 6-9. 2000. SAE 2000-01-1054.

Sellgren U. 2003. *Simulations in product realization - a methodology state of the art report.* Technical report, Dept. of Machine Design, Royal Inst. of Technology, 2003. ISRN KTH/MMK/R-03/05-SE. ISSN 1400-1179.

Sellgren U. & Hakelius C. 1996. A survey of PDM implementation projects in selected Swedish industries. *ASME Design Engineering Technical Conference,* August 18-22, 1996, Irvine, California 1996.

Senge P. 1990. *The Fifth Discipline: The Art and Practice of the Learning Organization.* Currency Doubleday. ISBN 0-385-26095-4. 1990.

Sha L., Rajkumar R. & Gagliardi M. 1996. Evolving Dependable Real-time Systems. *Proceedings of the 1996 IEEE Aerospace Applications Conference.* Aspen, CO, February 1996. New York, NY: IEEE Computer Society Press.

Sharman D. & Yassine A. 2004. Characterizing complex product architectures. *Systems Engineering,* vol. 7, no. 1, p35-60.

Shaw M. 1989. Larger scale systems require higher-level abstractions. *ACM Sigsoft Software engineering notes.* Vol 14. No 3. p143-146.

Shchedrovitsky GP. 1966. Methodological problems of system research. *General Systems* 11.

Shigematsu T. 2002. Software Quality Management Applied to Automotive Embedded Systems. *SAE Convergence 2002.* Transportation Electronics. Detroit, MI. October 21-23. 2002. SAE 2002-21-0017.

Shimizu K. & Dill D. 2002. Using Formal Specifications for Functional Validation of Hardware Designs. *IEEE Design & Test of Computers.* July/August 2002. p96-106.

Siegers R. 2005. The ABCs of AFs: Understanding Architecture Frameworks. *Proc. INCOSE International Symposium 2005*. Rochester, NY. July 10-15. 2005.

Skyttner L. 2001. *General Systems Theory: Ideas and Applications*. World Scientific Publishing Co. Singapore. ISBN 981-02-4175-5. 2001.

Spreng M. 2002. iDrive – The New Interaction Concept for Automotive Cockpits. *SAE Convergence 2002*. Transportation Electronics. Detroit, MI. October 21-23. 2002. SAE 2002-21-0042.

SPSS. 2005. http://www.spss.com/ accessed 2005-09-01.

Steiner P & Schmidt F. 2001. Anforderungen und Architektur zukünftiger Karosserieelektroniksysteme. *10 Internationaler Kongress, Elektronik im Kraftfahrzeug*. Kongresshaus Baden-Baden. 27-28 September, 2001.

Stoll Ulrich. 2001. Sensotronic Brake Control (SBC) – The electro-hydraulic brake from Mercedes-Benz. *10 Internationaler Kongress, Elektronik im Kraftfahrzeug*. Kongresshaus Baden-Baden. 27-28 September, 2001.

Storey N. 1996. *Safety-Critical Computer Systems*. Addison-Wesley. ISBN 0-201-42787-7.

Stuecka, R. 2003. Bridging the Gap is not Enough – Life-cycle Management for Automotive Electronics and Software. In: *Global Automotive Manufacturing and Technology 2003*, Business Briefings Ltd, Cardinal Tower, 12 Farringdon Road, London

Sveiby, K-E. 1996. Transfer of knowledge and the information processing professions, *European Management Journal*, Vol. 14, No 4 August, 379-388.

SysML. 2005. *SysML Specification v. 0.9 Draft* - http://www.sysml.org/ - accessed April 2005.

Systemite. 2004 Internet site accessed February 2004: www.systemite.com

Talbi T. Meyer B. Stapf E. "A metric framework for object-oriented development. Proc. 39th International Conference and Exhibition on Technology of Object-Oriented Languages and Systems.", TOOLS 39. 29 July-3 Aug. 2001 Pages:164 – 172. 2001.

Teepe G, Remboski D & Baker R. 2002. Towards Information Centric Automotive System Architectures. *SAE Convergence 2002*. Transportation Electronics. Detroit, MI. October 21-23. 2002. SAE 2002-21-0057.

Thane H. 1997. *Safety and Reliability of Software in Embedded Control Systems*. Lic Thesis, Department of Machine Design, KTH.

Thiel S. & Hein A. 2002. Modeling and Using Product Line Variability in Automotive Systems. *IEEE Software*. No 4. July/August 2002. p66-72.

Topp K & Weber J. 2001. Information Technology – A Challenge for Automotive Electronics. *SAE 2001 World Congress*. Detroit, MI. March 5-8. 2001. SAE 2001-01-0029.

Totel E, Blanquart J-P, Deswarte Y& Powell D. 1998. Supporting multiple levels of criticality. *Twenty-Eighth Annual International Symposium on Fault-Tolerant Computing*. 23-25 Jun 1998. Digest of Papers. p70-79.

Toyota. 2005a. *Concept Cars Personal Mobility*. Internet reference, http://www.toyota-europe.com/design/concept_cars/pm/ accessed 2005-05-16.

Toyota. 2005b. *Toyota - Prius*. Internet reference, http://www.toyota-europe.com/showroom/Prius/kce_4.html accessed 2005-05-16.

TTTech 2005. *TTTech - Time-triggered technology*. http://www.tttech.com/technology/

Törngren M. 1998. Fundamentals of implementing Real-time Control applications in Distributed Computer Systems. *Journal of Real-time Systems*. Vol 14. p219-250. Kluwer Academic Publishers.

Törngren M, Eriksson B, Wikander J & Vågstedt N-G. 2001. *Cost and Dependability in X-by-wire systems: Research Proposal*. Vinnova diary number 2001-05571.

Törngren M. & Larses O. 2004. *Characterization of model based development of embedded control systems from a mechatronic perspective - drivers, processes, technology and their maturity.* Technical Report TRITA-MMK 2004:23 ISSN 1400-1179. Royal Institute of Technology, KTH, Stockholm, 2004.

Törngren M & Larses O. 2005. Maturity of model driven engineering for embedded control systems from a Mechatronic perspective. In: *Model Driven Engineering for Distributed Real-time Embedded Systems*. Sébastien Gérard, Jean-Philippe Babau, Joel Champeau (editors). ISBN: 1905209320. August 2005.

Ulrich K. 1995. The role of product architecture in the manufacturing firm. *Research Policy*, vol. 24, no. 3, p419-440.

Vanderbilt. 2004. "GME: The Generic Modeling Environment." Internet site accessed February 2004: http://www.isis.vanderbilt.edu/projects/gme/index.html

Volvo. 2001. *Sirius brochure*. http://www.cad.luth.se/education/coursedocuments/Sirius2001_eng.pdf Downloaded 2002-01-18.

Wagner G. 2003. *Transmission options*. SAE Automotive Engineering Online. www.sae.org/automag/features/transopt/ accessed 2003-01-20.

Watt G. 2000. Firewalls in Safety-Critical Software Systems. *Proceedings of the 18th ISSC*. Sept. 2000.

Webopedia. 2003. Telematics. *Webopedia.com* www.webopedia.com/TERM/t/telematics.html accessed 2003-01-17.

Weber M. & Weisbrod J. 2003. Requirements Engineering in Automotive Development: Experiences and Challenges. *IEEE Software*. January/February 2003. p16-24.

Weinberg, G. 2001. *An introduction to general systems thinking (silver anniversary ed.)*, Dorset House Publishing Co., Inc., New York, NY, 2001

Wikander J., Törngren M. & Hanson M. 2001. Mechatronics Engineering - Science and Education, Invited Paper. *IEEE Robotics and Automation Magazine*, Vol 8, No. 2, 2001.

Wikipedia. 2005. Wikipedia, the free encyclopedia. http://en.wikipedia.org/wiki/Main_Page accessed 2005-11-07.

Wilson D., Murphy B. & Spainhower L. 2002. Process on Defining Standardized Classes for Comparing the Dependability of Computer Systems. *International Conference on Dependable Systems and Networks. DSN 2002*. Washington D.C. USA June 23-26, 2002.

Winter D. 2002. Open Systems Architecture – A Boeing Product Line Strategy for Avionics Systems. *SAE Convergence 2002*. Transportation Electronics. Detroit, MI. October 21-23. 2002. SAE 2002-21-0058.

Wolf W. 2002. Household hints for Embedded System Designers. IEEE Computer, May, 2002.

Würtenberger M. 2002. Business and Technical Impact of Open and Proprietary Architectures on Infotronic Systems. *SAE Convergence 2002*. Transportation Electronics. Detroit, MI. October 21-23. 2002. SAE 2002-21-0048.

X-by-wire. 1998. *Safety related fault tolerant systems in vehicles*. Final report. Project no BE95/1329. BriteEuRam III.

Yokohama T, Imoto Y & Takeshita T. 2002. Feasibility Study of High Speed Wheel Torque Control and its Effects on Vehicle Dynamics. *SAE Convergence 2002*. Transportation Electronics. Detroit, MI. October 21-23. 2002. SAE 2002-21-0026.

Yourdon, E. 1993. Yourdon™ Systems Method: Model-Driven Systems Development. Prentice-Hall, Englewood Cliffs, NJ.

Zimmerman, T. 2005. *Information Management for Mechatronic Products*. Licentiate thesis. Chalmers (2005)

Öberg T. 1998. *Modulation, detektion och kodning*. Studentlitteratur. Lund. 1998.

Acronyms

AADL – Architecture Analysis and Design Language (SAE standard)
ADL – Architecture Description Language
AMT – Automated Manual Transmission
AP – Application protocol (in STEP)
ARP – Average Ratio of Potential
CAx – Computer Aided X (anything)
CAD – Computer Aided Design
CAE – Computer Aided Engineering
CACE – Computer Aided Control Engineering
CASE – Computer Aided Software Engineering
CAI – Computer Aided Integration
CAN – Controller Area Network
CMM – Capability Maturity Model
CMMI – Capability Maturity Model Integration
COTS – Commercial Off The Shelf (product)
DoDAF – Department of Defence Architecture Framework
DSM – Design Structure Matrix
ECS – Embedded Control Systems
ECU – Electronic Control Unit
EE – Electrical/Electronic
EFFBD – Enhanced Functional Flow Block Diagram
GST – General System Theory
HMI – Human Machine Interface
INCOSE – International Council of Systems Engineering
MBD – Model Based Development
MBD – Model Based Development: Synonym of MDE – same definition.
MDE – Model Driven Engineering.
MI – Modular Independence
OEM – Original Equipment Maufacturer
PDM – Product Data Management
RTOS – Real-Time Operating System
SCM – Software configuration management
SDL – Specification and Description Language
STEP – STandard for the Exchange of Product data
SysML – Systems Modelling Language
TTP – Time Triggered Protocol
UML – Unified Modelling Language
VOR – Vehicle off road

Case study acronyms:
ACC – Adaptive Cruise Control
CTA – Complete Truck Architecture
PTA – Partial Truck Architecture
SAINT – Self Adaptive Intelligent Truck
FDoc – Function Documentation

Definitions

Accident. An accident is an unintended event or sequence of events that causes death, injury, environmental or material damage. [Storey 1996]

Accident prevention. Accident prevention is also referred to as active safety. The purpose of the measures is to avoid accidents.

Analytically redundant. Analytically redundant systems are allowed to produce different outputs if all the results comply with a defined model, the systems are said to show well formed diversity.

Assigning interface. The mapping of functional blocks to the physical constituents.

Availability. The availability of a system is the probability that the system will be functioning correctly at any given time. [Storey 1996]

Common cost. Common costs can not be traced to a specific product or service.

Cost. Cost is consumption of resources.

Damage prevention. Damage prevention is also known as passive safety and aims at reducing the consequence of an accident.

Dependability. Dependability is a property of a system that justifies placing one's reliance on it. [Storey 1996]

Development process. The development process entails all activities from new ideas to the point where changes are introduced in the production of the product.

Direct cost. Separate costs that are easily measured.

Embedded control system. A computer system that is part of a larger system and performs some of the requirements of that system; for example, a computer system used in an aircraft or rapid transit system.

Environment. The combination of conditions or constraints that determine the existence, and consequences of a system.

Error. An error is a deviation from the required operation of the system or subsystem. (Storey 1996)

External hazard. An external hazard is a situation, caused by events outside the system specification or due to a faulty or incorrect specification, in which there is actual or potential danger to people or to the environment

External safety. External safety is freedom from accidents or losses caused by events outside the system specification or due to a faulty or incorrect specification.

Fail-safe. In a fail-safe system a defined safe state exists, and it is always possible to enter this safe state in case of a fault or failure.

Fail-silent. A fail-silent system always responds to faults by omitting the output.

Failure. A system failure occurs when the system fails to perform its required function. (Storey 1996)

Fault. A fault is a defect within the system. [Storey 1996]

Fixed cost. Fixed costs are independent of production volume.

Function. The functions and logics of a system that interpret what is required as intents and qualities and define a solution by structuring the functions.

Functionally redundant. Functionally redundant systems produce identical output based on the same input, but allow internal diversity.

Graceful degradation. A system owns the graceful degradation property if a fault does not cause system failure but reduced system functionality.

Hazard. A hazard is a situation in which there is actual or potential danger to people or to the environment. [Storey 1996]

Implementation. The collection of physical parts of a system that perform, carry out, or realize what is required of the system.

Incident. An incident is an unintended event or sequence of events that does not result in loss, but, under different circumstances has the potential to do so. [Storey 1996]

Indirect cost. All common costs, and the separate costs that are not easily measured by product.

Information model. An information model is a semantic map where informational concepts and their relations are defined.

Integrity. Non-occurrence of improper alterations of information leads to integrity.

Internal hazard. An internal hazard is a fault that can lead to a system failure in which there is actual or potential danger to people or to the environment

Internal safety. Internal safety is freedom from accidents or losses caused by system failure.

Interpreting interface. A selection of the environmental issues that are of particular concern in determining functions and behavior logics.

Meta. (Greek: "about," "beyond") Meta is a common English prefix, used to indicate a concept which is an abstraction from another concept, used to analyze the latter. For example "metaphysics" refers to things beyond physics, and "meta language" refers to a type of language or system which describes language.

Model. A model is a simplified representation of a real or imagined system that brings out the essential nature of this system with respect to one or more explicit purposes.

Model based development (MBD). Development based on abstract representations with predefined and documented syntax and semantics, supported by tools.

Observer. An observer is a human outside a system observing the system. A system may serve a purpose, providing a function in the eyes of an observer.

Reliability. Reliability is the probability of a component, or system, functioning correctly over a given period of time under a given set of operating conditions. [Storey 1996]

Replication. Replication is a mere duplication of a system.

Resolving interface. Constraints and physical interactions between the constituents and the environment.

Risk. Risk is a combination of the frequency or probability of a specified hazardous event, and its consequence. [Storey 1996]

Safety. Safety is freedom from accidents or losses. [Leveson 1995]

Sales-to-delivery process. The sales-to-delivery process is initiated in the contact with a customer choosing a specific configuration, linked by production and ended when both product and sufficient knowledge about the product have been transferred to the customer.

Separate cost. Separate costs are costs that are traced to a specific product or service.

Service process. The service process entails all interaction with owners and users of the vehicle after the delivery.

Single point of failure. A system has a single point of failure if a single fault in a component can cause a complete system failure.

System. A system is a bounded set of observed objects with relationships between the objects and their attributes, selected by an observer from a wider set of objects with relationships between the objects and their attributes.

Variable cost. A variable cost depends on the production volume of the product.

"Är det så?"

"Så är det!"

To Ingrid and Viggo

Wissenschaftlicher Buchverlag bietet

kostenfreie

Publikation

von

wissenschaftlichen Arbeiten

Diplomarbeiten, Magisterarbeiten, Master und Bachelor Theses
sowie Dissertationen, Habilitationen und wissenschaftliche Monographien

Sie verfügen über eine wissenschaftliche Abschlußarbeit zu aktuellen oder zeitlosen
Fragestellungen, die hohen inhaltlichen und formalen Ansprüchen genügt,
und haben **Interesse an einer honorarvergüteten Publikation**?

Dann senden Sie bitte erste Informationen über Ihre Arbeit per Email
an info@vdm-verlag.de. Unser Außenlektorat meldet sich umgehend bei Ihnen.

VDM Verlag Dr. Müller Aktiengesellschaft & Co. KG
Dudweiler Landstraße 125a
D - 66123 Saarbrücken

www.vdm-verlag.de